Gender, AIDS and food security

Gender, AIDS and food security

Culture and vulnerability in rural Côte d'Ivoire

AWLAE SERIES No. 11

Mariame Maiga

Wageningen Academic
Publishers

ISBN: 978-90-8686-159-0
e-ISBN: 978-90-8686-715-8
DOI: 10.3921/978-90-8686-715-8

ISSN: 1574-2474

Cover photo: Leontine Visser, 2007

First published, 2010

This thesis is dedicated to the memories of my heroes

my sweet dad, Zakaria Maiga
my sweet mum, Djenebou Maiga Maiga

who have devoted all their life to assist orphans and vulnerable people

Acknowledgements

Coming from a community where girl's education is not a priority, doing a PhD was not only an ultimate dream, but also a personal commitment to challenge cultural barriers and gender disparities in women's education in developing countries. I am highly indebted to several persons and institutions without whose support it would not have been possible to accomplish my PhD studies.

I am thankful to the Netherlands government who funded my PhD studies, through the African Women Leaders in Agriculture and Environment (AWLAE) program of Winrock International. Sincere thanks to the leaders of the AWLAE PhD program at Nairobi and Wageningen: Professor Anke Niehof, Dr. Lisa Price, the late Professor Julia Gitobu, and Ms. Norah Muthike. Many thanks to Ms. Hedy Munro and Ms. Margaret van Wissen for their hard work for this exceptional and huge program, and for being so supportive to me and always willing to help.

I am indebted to the Anne van den Ban foundation for their financial support and to UNU-WIDER at Helsinki for an internship, from which I learned a lot. I am grateful to my internship supervisor Dr. Amelia Santos Paulino, to the Director, Deputy Directors, research fellows and the staff for making this internship an invaluable experience.

My special and profound gratitude goes to my supervisors, Professor Leontine Visser and Professor Anke Niehof. I learned so much from their leadership skills, the strong commitment to their students' work, and their humanism that allows one to take into account both the students' academic and social needs. Dear Profs, you have been such a source of motivation for me, not only for this PhD program but also for other achievements in my life. I would like to thank Mariette van Staveren for the English editing, Roy Jordaan for critically going through the manuscript, Mike Jacobs and Marijn van der Gaag for the work put into the production of this book.

Dr Kwikkel, I cannot find enough words to express my gratitude for taking care of my health. Steven and Monica, Kadi Ouattara and Laurie Goodman, many thanks for your support.

To the staff and colleagues of Rural Development Sociology Group, I am grateful for the friendship we shared. Special thanks goes to Jos Michel for being so supportive and always making sure I was comfortable in the group, to my brother Katani and my sister Lilian with whom I shared the office, and to the yoga team Carolina and Laura. Katani, thank you so much for your assistance with the quantitative data analysis.

The AWLAE PhD program allowed me to strengthen my academic and professional career and made me part of the network of AWLAE sisters: Fatimata, Aifa, Namizata, Kidist, Susana, Stephanie, Ekaete, Faith, Rose, Joyce, Monica, Gaynor, Regina, Lydia, Corrie, Hirut, Doris, and Carolyne. Although I do not have children of my own, the AWLAE babies Bineta Sow and Awa Niane made me a proud mother. Oumar Sow, Bineta, Mariame, and Kady from Belgium, Nana, Assane, Jean-Claude, Alfred, Idrissou, Petronella, Michel, Adinda, Juanwen, Zida, Djack, Amadou, Esther, Melissa, and Hom, many thanks for your friendship during my stay at WUR. Tata Oumy Khairy, thank you for your support, and Doortje Wartena, thank you for your hospitality. Many thanks to Gemma, Annelis, Hilde, Lucie, and Mieke of DIS. I will never forget the Denmark trip that Ellen so generously undertook with me.

My sincere thanks go to my English teachers Loukou Gerard and Yao for their commitment during my training in the American Corner Centre in Abidjan.

I express my sincere appreciation to the staff of AATF, Nairobi, and particularly the board of trustees that I am part of, for their support to my work and for making my joining the board a success.

My sincere gratitude goes to all those who assisted me in the research process and made the fieldwork possible: Dr. Diabaté Joséphine, Dr. Dieneba Coulibaly-Traoré, Agnero Raphael, Dr. Touré, Dr. N'cho, Dr. Kassoum Traoré, Babalasso, Sam, Kakoutié, Paterne, Dr Aka, and the whole research team. Dr Kassoum Traoré and Babalasso, I am highly indebted to you for the special and great contribution you made in this thesis. I thank you so much for your commitment and non-stop support to fill the gaps during the writing up of the thesis. I express my sincere appreciation to Kaddy Kamaté, Bintou Yara, Dr Doumbia Diénéba, Nadine Assémien, Djedji Cathérine, fatou Koné, Emma Assamoi, Mariame Traoré for their love and support I never missed.

My special thanks go to Issa Sanogo, papa Moussa Sanogo and their family, for housing me during the fieldwork. I thank you for giving me a second family.

I am highly indebted to my sisters, brothers and cousins Alima, Baba, Bah Cissé, Baroh, Tenin, Ben, Alassane, Jean, Mina, Fatoumata, Abdoul, Aida, Kara, Nass, Ousmane, Sali, Kouroundou, Oumou, Modibo, Balobo, Oumar Fofana, Adiza, Saydou, Abou, Mariame Kima, Béré, Astou, Bako, Adja, Amina, Awa Ouédraogo, Ami, Madjara, Oumou and Mariame Diop, Alama, Fanta, Bintou, Kadia, Odile, Mariame Tembély, and Fousseni for their support.

Special thanks go to Mamadou Cissé, without whom I would not have been able to go to the university, and to Astou Diabaté, Fadima Keita, Kady Korsaga for supporting my schooling.

I express my sincere appreciation to Check Cherif Oumar Aziz, Ibrahim Zon, Massamba, and Abdoulaye for their spiritual support which made me confident during this long journey. Clara De Sousa, you remain a role model for me.

Last but not the least, my very special gratitude goes to Ibrahim Magazi, Ibrahim Sy Savané, Isidore Boutchué and Phanse Mariko, who believed in me and never doubted that I could make it.

Besides my dad and my mum, I dedicate this thesis to my adorable niece Oumou Cathérine Traoré.

Mariame Maiga

Table of contents

List of tables

List of figures

List of acronyms

ANADER	National Agency for Rural Development
ARV	Anti Retro Viral
CBO	Community Based Organization
CMEC	Mutual Saving and Credit
DFID	Department for International Development
EGPAF	The Elizabeth Glaser Pediatric AIDS foundation
FAO	Food and Agriculture Organization
FDG (s)	Focus Group Discussion(s)
FENACOVICI	National Network of Food Crop Production Cooperatives
GDI	Gender Development Index
GDP	Gross Domestic Product
HBC	Home Based Care
HH	Household
HIPC	Highly Indebted Poor Countries
HIV/AIDS	Human-Immune-Deficiency Virus/Acquired Immune Deficiency Syndrome
IMF	International Monetary Fund
MOA	Ministry of Agriculture
MOPH	Ministry of Public Health
MORA	Ministry of Response to AIDS
MOHIV/AIDS	Ministry of the Fight Against AIDS
MOE	Ministry of Environment
NGO	Non Government Organization
OIPR	National office of parks and reserves
OVC	Orphan and Vulnerable Children
PEPFAR	The United States President's Emergency Plans for AIDS relief
PIPs	Policies Institutions and Processes
PMTCT	Prevention of Mother to Child Transmission
SODEFOR	National Institute of Forestry
SPSS	Statistical Package for Social Sciences
STD	Sexually Transmitted Disease
STIs	Sexually Transmitted Infections
UNAIDS	The Joint United Nations Programme on HIV/AIDS
UNDP	United Nations Development Program
UNFPA	United Nations Population Fund
UNICEF	United Nations Children's Education Fund
VTC	Voluntary Testing and Counseling
WB	World Bank
WCED	World Commission on Environment and Development

Glossary

Agnibougou	Street of the houses of Agni
Akpessi	Boiled or smoked banana, yam, cassava
Annayah	Third Friday of the moon
Annakissié	Third Monday of the moon
Awongnalé	Local name of disease
Baka	Local name of porridge made from maize or rice or millet
Babasso	Local name of disease similar to HIV/AIDS
Bêtê-bêtê	Soft yam varieties of *Dioscorea rotunda*
Bomoh	Traditional cloth used as gift for the dead
Bandji	Local name of drink made from palm oil tree
Chaude pisse	Sexually transmitted diseases (STD)
Comoé	Local name of river
Dedié	Sexually transmitted diseases (STD)
Djoulabougou	Street of the houses of migrants
Djoulaya	Business activities
Domake	Wage agricultural worker
Ehnwouhyaya	Long and persistent disease
Foutou	Local name of food made from pounded banana or yam or cassava
Itere	Name of a disease type of chronic malaria
kabato	Local name of food made from maize or millet flower
Kilié	Totem
Kabagnignin	Local name of food made from maize
Kokohndehn	Local name of food made from cassava flower
Komian	Local name of traditional healer
Maitresses	Female extra-marital partner
Malo	Local name of cooked rice
Malossainin	Local name of food made from cooked gratin of rice
Mna'	Holy Wednesday
Moumoumin	Traditional and witch dance
N'za	Yam varieties of *Dioscorea rotunda*
Tano	Local name of a river
Sohnsoribougou	Local name of township
Wafoutché	Short disease
Wayoawougnalé	Long and persistent disease

Chapter 1.
Introduction

This chapter introduces the rationale of this thesis, and its objectives. It presents the motivation why this research was undertaken and presents the research problem. The chapter concludes with the structure of the thesis.

1.1 Motivation for the study: why is it so difficult to defeat AIDS?

Reversing the AIDS epidemic remains definitely one of the major challenges of the 21st century. In 2000, the United Nations Security Council acknowledged AIDS as primarily a health and development issue that affects the developing world the most. The international attention led to an increased visibility of the epidemic through a strong mobilization of political, financial and human resources to defeat it, especially in Africa. However, UNAIDS (2008a) agrees that the global AIDS epidemic is continuing to challenge the worldwide efforts to be reversed, regardless of the mainstream knowledge, policies and interventions, along with the impressive financial resources allocated to challenge the disease in developing countries. According to this report, Sub-Saharan African countries remain heavily affected by the epidemic. 'An estimated 1.9 million (1.6 million-2.1 million) people were newly infected with HIV in 2007, bringing the number of people living with HIV to 22 million (20.5 million - 23.6 million). Two thirds (67%) of the global total of 32.9 million (30.3 million-36.1 million) people with HIV live in this region, and three quarters (75%) of all AIDS deaths in 2007 occurred there' (UNAIDS, 2008a: 39). Women remain the hardest hit by the disease because they are more susceptible to HIV infection than men are, while they are also more vulnerable to the social and economic impacts of the epidemic. It is noteworthy that, in developing countries, women play a key role in household food security, since they are the main food producers and food providers. The AIDS epidemic affects their labour asset, and increases their burden as caregivers. In the context of the epidemic, the heavy reliance on female agricultural labour leads to both an on-farm and an off-farm income decline, and exacerbates food insecurity at the household, national and regional levels. In short, gender inequalities remain at the heart of the AIDS epidemic and its impacts, and require a gender-based paradigm to be applied to agricultural and rural development.

My motivation to do this research and write this PhD thesis stems from my desire to strengthen the mainstream knowledge, policies and interventions to respond to AIDS in sub-Saharan Africa, and particularly in Côte d'Ivoire. In this thesis, I address three important issues:

- The question of whether the mainstream knowledge and policies are appropriate in the African context. People tend to interpret AIDS as primarily a health issue, and neglect the socio-cultural dynamic that stresses women's vulnerability to the disease and its impacts on food security, the rural economy and social development.

- The lack of a gendered framework for research on livelihoods and food security in a situation in which an AIDS epidemic rages in Sub-Saharan African countries like Côte d'Ivoire.
- The need to document factors that increase women's social and economic vulnerability, especially in relation to the AIDS epidemic and food insecurity, and to highlight their agency in the way they cope with these circumstances.

1.2 The research problem

The UNAIDS 2008 report on the global AIDS epidemic acknowledged that Sub-Saharan Africa did not make any real progress in its response to AIDS, compared to the other regions of the world. Considerable proportions of persons in the economically productive ages continue to be affected, with women having higher prevalence rates than men have. Women are most affected, because they are the most susceptible to HIV and more vulnerable to AIDS impacts on agricultural and socio-economic development and on food and nutrition security. Therefore, it is believed that the disease increases hunger and jeopardizes efforts to achieve the Millennium Development Goals.

The understanding of AIDS as a medical condition is highly influencing the response to the epidemic. 'As a result, medical actors, like doctors, medical institutions, and the pharmaceutical industry, have traditionally played – and still play – an essential role in how the disease is understood and tackled' (Nauta, 2008: 2). Consequently, the anti-retroviral therapy and the use of condoms have been promoted worldwide as the basic means to tackle the epidemic. Therefore, 'western' medicine and health promotion have been imposed on developing countries, through programmes such as those of the WHO, UNAIDS and PEPFAR (Nauta, 2008).

According to Nauta, by focusing first and foremost on dominant biomedical and behavioural approaches, the complex social and structural factors playing a role in the spread and impact of AIDS in Africa have not been addressed. People's vulnerability, caused by inequalities within and between societies, shapes the spread and impacts of the disease throughout the continent. Women suffer more from the disease as 'it is the poor, the marginalized and the disadvantaged – in all societies – that suffer and succumb to HIV/AIDS' (Nauta, 2008: 2). In the same vein, Schoepf (2001) argues that the spread of AIDS and its mitigation policies in Africa are shaped by the cultural politics, i.e. the power relations between Western countries and African countries. There is a focus on international and global biomedical AIDS policies, rather than on the endogenous factors that may shape the spread of the disease and its impacts in Africa.

I concur with Nauda and Schoepf that with the mainstream explanations and the existing medication policy, we cannot tackle AIDS efficiently in African countries, like Côte d'Ivoire. There is a need to go beyond the development industry driven by structural economic factors, which makes it difficult to mitigate the disease on the continent. In Côte d'Ivoire, 59% of the population lives in rural areas and is involved in food production in one way or another. Women are the main food producers. Yet, little is known about women's vulnerability under culturally different conditions. Although the impacts of AIDS on food security have been well-documented (Rugalema, 1999; Tapouzis, 2000; Mutangadura, 2000; Haddad and Gillespie, 2001; Barnett and Whiteside, 2002; Loevensohn and Gillespie, 2003; De Waal and Tumushabe, 2003; Gillespie and Kadiyala, 2005; Müller, 2004, 2005a,b), relatively few studies have explicitly addressed socio-cultural and gender factors, such as the influence of culture and kinship systems on gender roles, women's sexuality, their vulnerability to AIDS impacts, and the consequences for food and livelihood security. Little is known about how cultural landscapes and kinship shape women's exposure to AIDS, and influence the way in which they respond to its impacts, i.e. how they cope with AIDS, economic hardship, and other adversity, such as forest degradation.

With a prevalence of 4.7% (IES, 2005; UNAIDS, 2008a), Côte d'Ivoire has one of the highest HIV/AIDS prevalence rates in West Africa, with a high rate of female infection (6.4% versus 2.9% for men). Mitigation policies and interventions are oriented towards preventing the spread of the virus, to treat infected people, and to provide care for those who have been affected by the epidemic. These combined interventions are not enough to reduce the wide impacts of AIDS in Côte d'Ivoire, as policies and interventions are not adequately addressing gender issues, which scale up the impacts of the epidemic on food security.

This thesis goes beyond the biomedical and socio-economic discourse that is dominant in the interpretation and mitigation of AIDS, to explore the gendered socio-cultural landscape and the more endogenous issues underlying the epidemic. It calls for a rethinking of AIDS mainstream knowledge, policies and interventions in sub-Saharan African countries like Côte d'Ivoire.

1.3 Research objectives and questions

1.3.1 Goals

This study aims to contribute to understanding the influence of socio-cultural dynamics of women's vulnerability to AIDS and the impacts on food and livelihood security, the role of women's agency in coping with internal and external vulnerability, and the factors that affect their resilience. The study calls for a gendered approach to AIDS mitigation policies and interventions. Its findings will be used to improve the mainstream forms of knowledge, policies and interventions to respond to AIDS, and to sustain food security in Côte d'Ivoire. The research was carried out in the district Y.F.[1], among both autochthonous matrilineal Agni and patrilineal migrant populations.

I.3.2 Objectives

To analyze the influence of cultural landscapes on women's vulnerability to AIDS and on food security.

To assess the role of women's agency in coping with the impacts of AIDS, economic hardship, environmental degradation and cultural features, which together increase women's vulnerability.

To assess the relevance of AIDS and food policies in Côte d'Ivoire.

I.3.3 Research questions and sub-questions

1. How do gender roles, land rights, and access to resources affect women's vulnerability to AIDS and livelihood and food insecurity?
   - How do matrilineal and patrilineal kinship systems expose women to AIDS and impact on food security?
   - How do households and especially women experience labour shortages and how does this affect their food and livelihood security?
2. What is the interplay between AIDS, economic hardship, culture, and forest degradation in the research area?
   - How people interpret forest degradation and how does it impact food security?
3. How can the national AIDS mitigation policies and interventions be improved, taking into account the results of this study?

I.4 The structure of the book

In the present chapter, this thesis starts with a general introduction, along with the motivation underlying the research.

Chapter 2 presents information on the country and the study area. Geographic, demographic, economic and ecological characteristics are presented, with a specific focus on the social structure and cultural features of the study area. Government responses to the AIDS epidemic as well as agricultural and food policies are discussed. The chapter concludes with a presentation of the vulnerability context related to AIDS, and of the forest degradation in the study area.

[1] The abbreviation Y.F. is used throughout this thesis as an alias for the study area, as the issue of AIDS is very sensitive for the inhabitants of the district.

Chapter 3 presents a literature review on livelihood and vulnerability approaches, as well as a discussion on the key concepts of household and gender, forest degradation, and kinship. The chapter concludes with the presentation of a conceptual framework, which postulates the interconnection of concepts.

Chapter 4 presents the study design and the research strategy, and the methods used to collect and analyze data. It also addresses methodological discrepancies and the ethical considerations encountered during the study.

Chapter 5 examines women's internal vulnerability and the ways in which women respond to vulnerability at a micro- and household level, given the entitlement failures and lack of access to certain assets and resources in their everyday world. The chapter also discusses cultural norms, practices and stigmatization shaped by matrilineal and patrilineal kinship as factors that increase women's vulnerability to AIDS, and that impact on food and livelihood security.

Chapter 6 presents the external vulnerability context of food security in terms of crop production and access to extension services, credit and markets. It discusses forest degradation, health and the AIDS situation; the lack of access to medical facilities that households and particularly women experience as external vulnerability. Additionally, the chapter examines narratives on AIDS, the biomedical treatment, the traditional healing practices, AIDS and witchcraft. The chapter ends with a discussion on the relations between women's status and children's malnutrition, as well as the impact of external vulnerability on social capital.

Chapter 7 introduces a theoretical discussion on human agency and social practices. It sets a general analytical framework to understand women's resilience and resistance, and the way they use their agency to challenge the hardship they are faced with. It analyzes the way in which women use their agency to configure rules and resources in the process of their resilience.

Chapter 8 summarizes the key findings of the study to answer the research questions formulated. Policy and interventions recommendations are presented to improve AIDS policies and programmes as well as agricultural and food policies, to respond to the AIDS epidemic and sustain food security in the country.

Chapter 2.
Country and study area profiles

This chapter starts with a description of Côte d'Ivoire, followed by a description of the study area. It presents the population, economy and ecology of the country, with specific emphasis on the social structure and cultural background of the study area. Additionally, the chapter provides details on the rationale and challenges of AIDS policy and interventions in the country, as well as on agricultural and food policies. The chapter concludes with a description of the vulnerability context, related to forest degradation and AIDS in the study area.

2.1 Country profile[2]

2.1.1 Geography, population and migration

Côte d'Ivoire is a West African country, situated on the Gulf of Guinea (Figure 2.1). The country is bounded by the Atlantic Ocean in the south, Liberia in the southeast, Guinea in the northwest, Burkina Faso and Mali in the north, and Ghana in the east. Côte d'Ivoire is divided into nineteen regions. The regions are divided further into 81 Departments. While Yamoussoukro is the political capital, Abidjan is the economic capital and administrative centre, and also the largest city. The population of Côte d'Ivoire is estimated at 20,677,000 inhabitants (World Population Data, 2008). Côte d'Ivoire is a multicultural country, with a lot of ethnic and religious diversity. There are more than 60 ethnic groups, coming from four larger ethnic groups: the Akan, Voltaiques, Kru, and Mande. People adhere to one of the three main religions; they adhere either to the Islam or Christianity, or are classified as Animists when they do not (DSRP, 2009:1). French is the official language in Côte d'Ivoire, while the most frequently spoken local languages are *Bambara* and *Baoule*. The monetary unit of the country is the Franc CFA.

In 2008, the migrant population in Côte d'Ivoire was estimated at 26% of the national total (DSRP, 2009: 1). The country's prosperity and economic stability up to the late 1990s attracted numerous workers from neighbouring countries (Fieloux, 1980). Migrants coming from Burkina Faso, estimated at more than one million, are mostly agricultural workers, who seek employment on the cocoa and coffee plantations. Several hundred thousand Malians, Senegalese, Ghanaians, Mauritanians and Guineans live in the country as well. Demographic and land pressures have motivated clashes in rural areas over land ownership between migrant agricultural workers who own land and autochthonous people.

---

[2] In this chapter, I especially made use of the Poverty Reduction Strategies Document of Côte d'Ivoire (DSRP, 2009), which provides the most recent and relevant data on the country's profile.

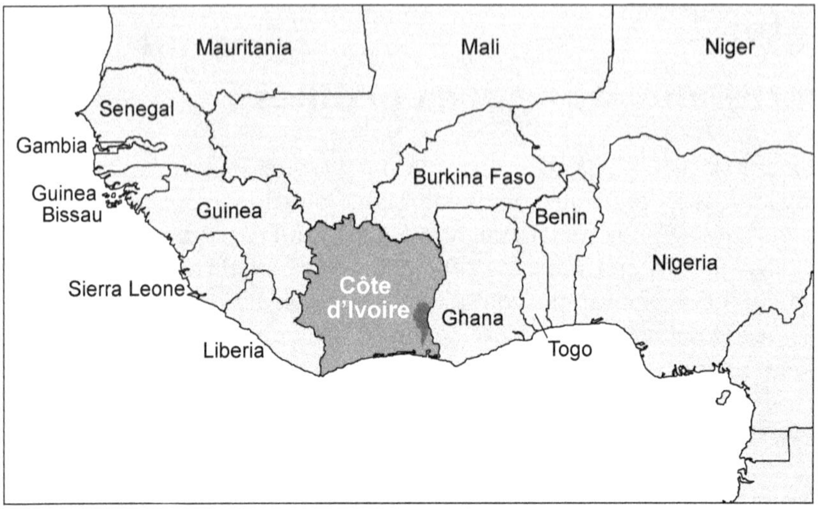

Figure 2.1. Regional map of West Africa: location of Côte d'Ivoire. The research area Y.F. is located within the district Abengourou (indicated dark grey).

2.1.2 Ecological background

Côte d'Ivoire has a tropical climate, with four seasons in the southern part: a long raining season from April to mid-July, a short dry season from mid-July to September, a short raining season from September to November, and from December to March, the long dry season. The northern part has two main seasons: a big raining season from June to September and a big dry season from October to May (Bourgouin and Guillaume, 1979). Côte d'Ivoire used to have a high level of biodiversity, with over 16,034 vegetal and animal species, including 163 mammals and 712 birds. The country has four big rivers: the Sassandra, Bandama, Comoé, and Cavally, and eleven basins. Like many developing countries, Côte d'Ivoire is faced with the depletion of its natural resources. Estimated at 12 million acres in 1960, the forest cover has dropped to 2.802 million acres in 2007, which means a forest loss of 75%. Agriculture, uncontrolled fires, and logging activities for tropical woods are the main causes of forest loss. In 2007, the reserved forest areas were estimated at 0.672 million acres, and the national parks and reserves areas were estimated at 1.728 million acres (DSRP, 2009: 63).

2.1.3 National, institutional and socio-economic context

Socio-political crisis and the end of a peaceful history

For three decades following its independence from France on 7 August 1960, Côte d'Ivoire has been the leading country in West Africa and a model in terms of economic and political stability in sub-Saharan Africa. Peace was like a second religion and Côte d'Ivoire has often been the refuge for those who fled from conflict in countries such as Liberia and Sierra Leone. However, starting in 1990, the country has been exposed to a socio-political crisis, with repeated *coups d'état*. One of these was the failed coup attempt of 19 September 2002, which has led to the country's worst civil war since its independence in 1960.

The crisis of 19 September 2002 has led to a rebellion and has split the country into two. The northern part is still being controlled by the rebels and the southern part by the government. National identity, eligibility for citizenship, and land tenure has been seen by many observers as among the root causes of the conflict. Up to 3,000 persons were killed during the civil war, and up to 700,000 persons were displaced. With the commitment of the international community, power-sharing agreements have been signed between the country's major political parties, the government and the rebels, to put an end to the civil war. The Linas-Marcoussis Accord (January 2003), the Accra III Agreement (30 July 2004), the Pretoria Agreement (6 April 2005), and the Ouagadougou Peace Accord signed on 4 March 2007 brought the concrete prospects of hope and of the end of the civil war (http://www.infoplease.com/country/profiles/cote-d-ivoire.html, accessed on 20-07-09). Côte d'Ivoire is now a fragile and post-crisis state. The civil war has jeopardized the national development agenda, impacted heavily on economic growth, caused the loss of foreign assistance and private investment, and increased the government's internal and external debt. The economic decline has worsened people's living standards, has caused repeated increases of the cost of food commodities, and has made it difficult for women and children to get access to basic social services, particularly to health care and food (ENV, 2008).

Socio-economic conditions

According to the World Bank, Côte d'Ivoire has the largest economy in the West African Economic and Monetary Union (WAEMU), accounting for 40% of its GDP. Its economic influence is still critical to the overall development of the West African region. With its large migrant population (26%), Côte d'Ivoire is an important source of worker remittances for countries in the region. The country's level of total goods' exports is the fourth largest in sub-Saharan Africa (after South Africa, Nigeria and Angola). The civil war has affected the country's economic growth and development. From 2000-2006, the average economic growth turned negative with -0.4%, and was below the WAEMU rate of 4.1% and the 4.9% rate in sub-Saharan Africa. (http://www.go.worldbank.org/SN2JJ08P10, accessed on 24-07-09). The Human Development Index (2006) for Côte d'Ivoire ranks 164th out of 177 countries.

The economy of Côte d'Ivoire depends heavily on the agricultural sector, and is largely export-oriented. Two-thirds of the economically active population is engaged in agricultural activities. The agricultural economy is diversified, and the country also has significant manufacturing and services sectors. Côte d'Ivoire is the world's leading producer of cocoa, and a significant producer of coffee and palm oil. The industrial sector performs in textiles, ship construction and repair, building materials, electricity, wood products, oil refining, fertilizers, foodstuffs, and beverages. Côte d'Ivoire has rich natural resources such as petroleum, diamonds, iron ore, hydropower, natural gas, manganese, cobalt, and copper. The country mainly exports cocoa, coffee, timber, cotton, palm oil, petroleum, bananas, pineapples and fish, while it imports foodstuffs, capital equipment, and fuel. The external debts of Côte d'Ivoire are estimated at almost four billion US$, i.e. one-third of the national budget, making it impossible to boost up the economic growth with investments (http://www.infoplease.com/country/profiles/cote-d-ivoire.html, accessed on 20-07-09).

The last national report on the socio-economic conditions of households in Côte d'Ivoire (ENV, 2008) reveals that almost half of the Ivorian population, i.e. 48.93%, is living on less than 1 euro (661 francs CFA) per day and per person. According to this report, the economic aftermath of the civil war in 2002 has increased people's socio-economic vulnerability. Unemployment has risen up to 40-50%. Most of the people have difficulties getting access to health care, due to poverty. Those who cannot afford modern medical treatment, rely on traditional medicine. The report shows (ENV, 2008) that rural populations remain definitely the most affected; 75% of the population who are poor, live in rural areas. Seen from a gender perspective, the report also shows that women remain poorer than men in rural Côte d'Ivoire.

In short, although Côte d'Ivoire has been admitted (27 March 2009) to the Heavily Indebted Poor Countries (HIPC) programme to benefit from financial support from the Bretton Woods institutions, the profound socio-economic crisis affecting the country is jeopardizing the achievement of the Millennium Development Goals at the national level. These goals are to reduce extreme poverty and hunger as well as female illiteracy, and child and maternal mortality by 2015.

The status of women and children

Poverty in Côte d'Ivoire is gendered. Women are significantly poorer than men. According to the UNDP (2007), the ratio of female-earned income to male-earned income for Côte d'Ivoire was 0.32 in 2005, while the GDP per capita figure of US$ 795 was lower for women than the national average of US$ 1.648. The gender-related development index value (GDI) for Côte d'Ivoire ranks 145th, with a value of 0.413. Like in many developing countries, there are gender disparities in education in Côte d'Ivoire. The male literacy rate is higher, with 70%, whereas the female literacy rate is 52% (PRB, 2008; Table 2.1).

Table 2.1. Main demographic indicators of Côte d'Ivoire (PRB, 2008; UNICEF, 2008).

Population (millions)	20,677,000
Population growth rate (%)	2.4
Life expectancy (years) M/F	50/53
% Literate >15-24 years M/F	70/52
Infant (ages 0-1) mortality rate (per 1000 live births)	90
Under-five mortality rate (per 1000 live births)	127
Maternal mortality ratio (per 100,000 live births)	810

In terms of political participation and decision-making power, women are not represented enough to influence policies that continuously jeopardize their emancipation, along with a sustainable environment and development. The country counts only three women ministers out of 32, and 19 women deputies out of 225.

The civil war has displaced thousands of people, particularly women and children. It has deepened the feminization of poverty, accordingly increasing women's vulnerability to HIV mainly in war-affected areas, where the rate of HIV infection remains much higher due to the widespread sexual violence and increased prostitution.

With the aftermath of the war, a considerable number of Ivorian women are still lacking the financial means to be healthy and economically secure, and thus are also unable to preserve their children's health. The lack of income leads to a lack of food and medical care for many women and their children. Seventeen percent of Ivorian infants have a low birth weight; the infant mortality rate and the under-5 mortality rate are estimated at 90 and 127 (per 1000 live births) respectively; 20% of children under the age of five are suffering from moderate or severe underweight (Table 2.1). The maternal mortality ratio of 810 is one of the highest rates world-wide (UNICEF, 2008). Hence, Côte d'Ivoire has some of the worst health indicators in Africa.

Agricultural policy and food security

The economy of Côte d'Ivoire is still heavily agricultural and market/export-based rather than food security-based. The main agricultural productions are coffee, cocoa beans, cotton, gum, cashew, sugarcane, cassava, bananas, corn, rice, palm kernels, sweet potatoes, yam, maize, and vegetables. The majority of the Ivorian population relies on smallholder cash-crop production. Yet, Ivorian women remain heavily marginalized in cash crop production because they don't usually have access to the resources (information, seed, and fertilizers) that are needed for it. They are engaged more in food crop production than cash crop production, providing the bulk of the food production (DSRP, 2009). However, the national agricultural policy neglects the food crop production sector as a means to attain food security, which would support women's agricultural activities. This is illustrated by the way in which the reforestation programme was implemented in the research area. According to the district's director of the national forestry agency, while the forestry agency was implementing the reforestation programme in the area, it has been asked to destroy any kind of food crops that were being produced in the reserved forests, which people were using for agricultural production. However, the destruction of cash crops such as cocoa and coffee was not listed in the programme as, according to the director, this would have had strong national economic impacts. The interest given to cash crops production at the detriment of food crops production portrays the national agricultural and food policies, and makes it difficult to respond in a sustainable way to the national demand in terms of food security. A national report from the Ministry of Agriculture shows that Côte d'Ivoire exports more than 50% of its rice to supply the national demand. The Central Emergency Response Funds, (CERF, 2008) has reported that Côte d'Ivoire faces a gradual deterioration of its national food security.

The recent global food crisis has raised the debate on the relevance of agricultural and food policies in developing countries like Côte d'Ivoire. While the exogenous factors of the global food crisis are related to the power relations in the international political economy, the endogenous factors relate to policy choices that favour macro cash crop production over food crop production, thereby neglecting the issue of food security.

2.3 The AIDS epidemic and AIDS policy in Côte d'Ivoire

Côte d'Ivoire has one of the highest HIV/AIDS prevalence rates in West Africa (Table 2.2). According to the national report (EIS, 2005), people living with HIV are estimated at 4.7%. Women are still the most hit by the epidemic, with an estimate of 6.4%, whereas it is 2.9% for men.

Table 2.2. HIV and AIDS in Côte d'Ivoire (UNAIDS, 2008a).

People living with HIV, 2007	480,000
Women (aged 15+) with HIV, 2007	250,000
Children with HIV, 2007	52,000
Adult HIV prevalence (%), 2007	3.9
AIDS death, 2007	38,000

Côte d'Ivoire is one of the rare countries that have created a Ministry exclusively dedicated to the Response to AIDS[3] (*Ministere de la Lutte contre le SIDA*). By creating this Ministry, the decision-makers were aiming at strengthening the national mitigation policies designed to challenge the HIV epidemic in the country, in collaboration with the Ministry of Public Health. Its duties include the raising of AIDS awareness, the mobilization of civil society and NGOs, advocacy, fundraising, and impact assessment of the epidemic on different economic and development activities, and the monitoring and evaluation of AIDS-related interventions. The mission of the Public Health Ministry relates to medical and therapeutic issues, like Anti-Retro-Viral (ARV) delivery programmes and the prevention of mother-to-child transmission (PMTCT). According to the current AIDS National Strategic Plan (NSP) for 2006-2010, the targets are to reduce HIV incidence by 25%, and to increase the number of HIV-infected people treated with ARVs from 20,000 to 104,000 by 2010, using a decentralized and multisectoral response to AIDS.

The national AIDS policy is monitored by the National Committee for the Response to AIDS (*Le Conseil National de Lutte contre le VIH/SIDA*), presided by the head of state, and composed of 140 local and international stakeholders involved in the response to the disease in the country (EIS, 2005). At the regional level, there are regional committees focusing on the response to AIDS, chaired by the *préfet de région*, whereas the *président du conseil général* and the mayor are the vice-presidents, and the regional director of health is the executive secretary. At the department level, there are similar committees, chaired by the *préfet*. At the community level, the committees are chaired by the *sous-préfet* and, at the village level, by the chief of the village. The regional committee is assisted by a Technical Support Unit. They have to design AIDS policies and strategies at the local level, based on a local needs assessment, whereas the technical staff of the AIDS Response Ministry has to implement these. In short, according to the new national AIDS policy framework, AIDS interventions are not performed and monitored exclusively by medical officers anymore, as they were in the past; now, civil servants and local communities are involved in this as well (EIS, 2005).

2.3.1 International agencies and programmes on AIDS

Similar to many other countries, Côte d'Ivoire has received assistance from international institutions, which are providing financial and technical support to defeat the epidemic in the country. These international agencies are PEPFAR, the Global Fund, the World Bank, WHO, UNICEF, UNFPA, UNAIDS, PAM, HCR, ONUCI, Care International, Hope Worldwide-C.I, the Elizabeth Glaser Pediatric

[3] Before the creation of the Response to AIDS Ministry on 24 January 2001, the response to AIDS was a programme of the Public Health Ministry, and afterwards became a programme of the Prime Ministry, coordinated by a delegated Ministry.

Aids Foundation (EGPAF), JHPIEGO/Johns Hopkins University, the International HIV/AIDS Alliance, Population Services International (PSI), the Association of Public Health Labs (APHL), the Projet Retroci-CI, Family Health International (FHI), and PSP Associates. Additionally, some AIDS programmes in the country rely on donations from bilateral development cooperation, particularly from Belgium, Canada, France, Germany, and Japan. Apart from the UN institutions, PEPFAR, the Global Fund, and the World Bank are the biggest sponsors of AIDS mitigation interventions in Côte d'Ivoire.

The United States President's Emergency Plan for AIDS Relief (PEPFAR)

Côte d'Ivoire is one of 15 focus countries of the United States President's Emergency Plan for AIDS Relief (PEPFAR). The eligibility of Côte d'Ivoire for the PEPFAR programme was possible because the country has one of the highest HIV prevalence in West Africa[4]. In Côte d'Ivoire, the PEPFAR supports AIDS prevention, treatment, and care programmes, with a particular interest in children orphaned by AIDS (http://www.pepfar.gov/press/81552.htm, accessed on 1-02 2008). Some stakeholders, believe that PEPFAR has a leadership position among the international AIDS agencies in Côte d'Ivoire, as the financial support provided by the programme is still increasing. In the implementation of its programmes, PEPFAR's main local partners are the Ministry of the Response to AIDS, the Ministry of Public Health, the Ministry of Woman, Family and Child, and the Ministry of Education. Additional partners are organizations that are active at the grassroots level, such as the National Institute of Rural Development (ANADER), Care International, Hope Worldwide-C.I, the Elizabeth Glaser Pediatric Aids Foundation (EGPAF), JHPIEGO/Johns Hopkins University, the International AIDS Alliance, Population Services International (PSI), the Association of Public Health Labs (APHL), the University of California-San Francisco, Supply Chain Management Systems (the partnership), the Projet Retroci-CI, Family Health International (FHI), PSP Associates, Repmasci, and Rip Plus. The Ministry of the Response to AIDS, the Ministry of Public Health, the Ministry of Woman, Family and Child, and the Ministry of Education each have a specific programme financed by PEPFAR, with regard to their mandate and goals. The Ministry of Woman, Family and Child is in charge of the National Programme for children orphaned by AIDS (COA); the Ministry of Public Health is in charge of the national programme for the palliative care/basic health care for people living with HIV, and also of the prevention of mother-to-child transmission (PMTCT) programmes; the Ministry

[4] The PEPFAR programme has been launched in over 100 countries around the world, with targeted actions in fifteen of the globe's worst-affected countries, including Côte d'Ivoire. PEPFAR's financial contribution to AIDS mitigation interventions in Côte d'Ivoire is estimated at US$ 404 million for 2006-2010. Therefore, compared to the funding release by the UN systems, US$ 56 million, the Global Fund, US$ 18 million and the started World Bank MAP, US$ 20 million, PEPFAR is the leading international agency, in terms of the magnitude of funding for HIV in Côte d'Ivoire (UNAIDS, 2008b).

of the Response to AIDS is in charge of AIDS awareness-raising programmes and advocacy and capacity building programmes. The Ministry of Education is in charge of the A (abstinence) and B (be faithful) parts of the ABC (condom) paradigm.

In sum, PEPFAR is providing both downstream and upstream support, with its financial and technical support to more than 100 partners and sub-partners, from Ministries to local communities and organizations in Côte d'Ivoire. Among the PEPFAR AIDS mitigation programmes, the prevention of mother-to-child-transmission (PMTCT) programme is lacking sufficient financial and technical support, given pregnant women's and children's vulnerability to the epidemic. In rural areas, the majority of women and their children are not benefiting enough from such programmes. The prevention of mother-to-child-transmission programme is not functioning adequately, as women cannot respect the medical protocols to visit the health center regularly, and to provide formula milk to their potentially infected babies. The urban-centered programme is still lacking a systematic approach that incorporates the socio-economic and socio-cultural vulnerability of women in rural areas, to protect babies from an HIV infection. The programme might be reinforced with the integration of voluntary family planning and maternal health services into the PEPFAR programmes.

The Global Fund to fight AIDS, tuberculosis and malaria

The Global Fund initiative is coordinated by the United Nations Development Programme (UNDP). The Global Fund is supporting programmes on AIDS awareness with an emphasis on community mobilization, counseling and testing, the prevention of mother-to-child-transmission programmes, and capacity building of national institutions. The main local partners of the Global Fund are the Ministry of the Response to AIDS, the Ministry of Public Health, and some organizations, such as the National Institute of Rural Development (ANADER), the Public Health Programme (PSP), and the National Programme of Counseling (PNPEC) who are working on the ground. Regarding its agenda, the Global Fund has implemented four regional AIDS mitigation programmes, and five sectoral AIDS mitigation programmes. The country grant portfolio, released by the Global Fund to fight AIDS, tuberculosis and malaria in Côte d'Ivoire is estimated at US$18 million for 2006-2010 (UNAIDS, 2008b). The Global Fund programme is as yet not operative in the research area.

The World Bank's Multi-country AIDS Programme (MAP)

Côte d'Ivoire is one of the countries that have benefited from the Multi-country AIDS Programme for Africa (MAP), launched by the World Bank to mitigate the epidemic in sub-Saharan Africa. The Multi-country AIDS Programme aims at supporting the national AIDS strategies and interventions, by scaling up prevention, care, and treatment programmes, capacity development, and by mitigating the impact in all sectors and strengthening the interventions. In Côte d'Ivoire, the MAP programme is implemented by the Ministry of the Response to AIDS, and supports community-based organizations, NGOs, and the private sector for local AIDS initiatives. The World Bank has released US$ 20 million for the implementation of the Emergency Multi-Sector HIV/AIDS Project in Côte d'Ivoire (http//go.worldbank.org/30S1D37550). However, this programme is as yet not operative in the research area.

2.3.2 AIDS policies and interventions and leadership problems

In the national AIDS policy framework, a Coordination Committee has been appointed to coordinate the overall AIDS interventions in the country. This committee is a large forum composed of the representatives from public and private institutions, international AIDS agencies, the Ministry of Public Health, the Ministry of Finance, civil society, and other development aid partners. Every three months, this committee holds a meeting with the Ministry of the Response to AIDS, to assess and improve AIDS mitigation activities. However, this mechanism is not working adequately, as a strong coordination is lacking at the national level.

The creation of an autonomous Ministry of the Response to AIDS, detached from the Public Health Ministry, led to a crisis of collaboration, caused by a problem of leadership between both ministries.

> *'Unfortunately, the relations between the Ministry of the Response to AIDS and the Ministry of Public Health seem to be conflictual. There is a leadership problem that hampers AIDS mitigation policy and interventions in Côte d'Ivoire. The Ministry of Public Health does not participate in the "World Aids Day", known to be the biggest event in AIDS awareness and response. The political situation has a serious impact on the national mitigation strategies, because the Ministries are staffed by officials from different political parties. Consequently, the Ministry of the Response to AIDS did not yet develop relevant mechanisms to coordinate AIDS intervention activities in the country' [A programme coordinator in an international AIDS agency, Abidjan, 24-05-09].*

The problem of leadership between both ministries affects somehow the mitigation process. On the ground, there is duplication of some AIDS mitigation projects as well as the lack of coordination of the interventions of international agencies. These agencies do not clearly collaborate; they do not inform each other before implementing mitigation projects, as they have to rely on their international agenda. The Ministry of the Response to AIDS complains about the international agencies such as PEPFAR, the Global Fund, and the UN institutions, which focus on their own agenda and not on the national and local agenda. The Response to AIDS Ministry is not involved in the monitoring of their interventions, either. These international agencies attribute the weaknesses in the coordination to a leadership problem and a lack of collaboration among the main local stakeholders, particularly between the Ministry of the Response to AIDS and the Ministry of Public Health.

In short, the creation of a Ministry of the Response to AIDS illustrates the motivation of Ivorian decision makers to efficiently challenge the epidemic. The framework of the national AIDS policy and its rationale may offer the potential of change, regarding the positioning of AIDS as not primarily a health problem, but also a socio-cultural problem, which would enable the design of appropriate mitigation strategies. The framework was shaped by a top-down approach that could ensure the involvement of national and international stakeholders, as well as that of village stakeholders. However, the clash between the main national stakeholders, i.e. the Ministry of the Response to AIDS and the Public Health Ministry, has reinforced the thought that the creation of a national response to AIDS was based more on political interest than on considerations of efficiency[5].

AIDS policies and interventions in Côte d'Ivoire suffer not only from a leadership problem between both ministries, but also from a leadership problem among the international institutions and NGOs operating in the mitigation context in the country. The leadership problem among international AIDS agencies illustrates the complexities of the response to the epidemic in developing countries like Côte d'Ivoire, as the interventions are undermined by power relations between the international agencies, local partners, and the people. According to Nauta (2008), new problems are emerging from the non-democratic way in which donors and multilateral organizations are delivering their services to respond to AIDS in Africa; 'Many health projects are executed by foreign NGOs and academic groups, operating with almost no government interference inside weak or failed states, leaving the poor with virtually no provisions to say what they want [and] decide which projects serve their needs, or adopt local innovations' (Garret, 2007: 16). These problems cause the issues regarding the coordination of AIDS interventions that impact on the response to the epidemic in African countries. What is more, while the 'AIDS industry' is poaching local talents, it weakens the local health system and jeopardizes the sustainability of the response to the epidemic in vulnerable countries like Côte d'Ivoire.

[5] In Côte d'Ivoire, the creation in 2001 of a Ministry of Response to AIDS, detached from the Public Health Ministry, raised a debate about the real intentions of the government, as some people believe that an AIDS Response Ministry is a potentially rich Ministry. Besides, up to now, the Public Health Ministry has been ruled by the opposition, whereas the Response to AIDS Ministry has been ruled by a minister from the political party in power since its creation.

2.4 ANADER, agricultural extension services and the response to AIDS

In Côte d'Ivoire, the agricultural extension services are performed by the National Institute of Rural Development (ANADER), created in 1994. Its overall mission is to improve the national agricultural production in order to attain national food security and rural development. To achieve this ambitious goal, the agency operates in 57 zones, covering the national territory. ANADER services are both public and private. The agency provides private extension services for non-governmental agricultural programmes. Otherwise, the agency's services are said to be public and free of charge for farmers. The ANADER has extension officers, even in remote areas. These officers are supposed to assess the agricultural problems of farmers and provide them with the technical assistance and follow-up they may need to improve their production.

Two AIDS programmes, i.e. the PEPFAR-ANADER programme, and the Fond-Mondial-ANADER programme, are being implemented in rural areas. In the completion of its overall mission, ANADER has committed itself to the response to AIDS in rural areas, with the financial and technical support from international AIDS agencies like PEPFAR and the Global Fund. Focusing on its expertise in rural development and its presence in rural communities, ANADER has been appointed a key actor in the implementation of the national plan of the response to AIDS in rural Côte d'Ivoire. ANADER AIDS programmes carry out prevention, treatment and care in rural areas, in collaboration with the local health center. PEPFAR and Fonds-Mondial programmes have specific and different areas of intervention, although they have the same goal, i.e. to spread AIDS awareness and challenge it among rural communities in order to protect agricultural production. In the study area, the ANADER AIDS programme that is being implemented so far is the PEPFAR-ANADER programme.

2.4.1 The Fonds-Mondial-ANADER programme

The Fond-Mondial-ANADER programme so far implemented is operative in twelve districts[6]. The programme provides capacity building for NGOs involved in the response to AIDS, in order to strengthen their effectiveness when responding to the epidemic. The program is executed in partnership with local health centers, NGOs, and the local committees involved in AIDS mitigation activities. Unlike the PEPFAR-ANADER programme, the Fond-Mondial-ANADER programme includes off-farm companies workers in its targets population.

[6] These districts are Tiebissou, Toumodi, Yamoussoukro, Bouafle, Zuenoula, Sinfra, Grand-Lahou, Dabou, Sikensi, Tiassale, Divo, and Lakota.

2.4.2 The PEPFAR-ANADER programme

The implementation of the PEPFAR-ANADER programme required ANADER to collaborate with key networks and stakeholders involved in the national response to the epidemic, i.e. REPMASCI, a network of journalists, and ANACONDA, a network of medical scientists. The PEPFAR-ANADER programme is implemented in six regions and 24 villages[7]. Regarding its agenda, each village is entitled to sessions of sensitization and prevention, testing and counseling[8]. In a selected village, one day before the testing and counseling, actions of sensitization are organized by REPMASCI, through movies, music and dance performances on the AIDS epidemic and its impacts. The following day, the testing is performed by medical scientists, using a mobile unit. People who are tested HIV-positive get counseling, treatment and care. This is the way in which officers from the PEPFAR-ANADER programme and the local health center work together in testing and counseling. Treatment is the responsibility of the health center, whereas care is performed by the village community advisors, called *conseillers communautaire*, a team that belongs to the village committee of the response to AIDS, that is *comité villageois de lutte contre le SIDA*. People from this committee do the psycho-social care after the testing. They deliver care services in HIV-positive households, on a regular basis. Children orphaned by AIDS are entitled to a school kit and other related support from the programme. However, the actual practice may differ, as people in the villages often fear stigmatization.

2.5 Description of the study area: modernity and tradition

2.5.1 Location, social structure and socio-economic characteristics

The research area is a village located in Abengourou, a district belonging to the Moyen-Comoé region, in eastern Côte d'Ivoire. The district is situated at the border of Ghana. Abengourou is the central location for the Agni people. In the Agni language, Abengourou means 'I don't want trouble'. In 2010, the population of Abengourou is estimated at 110,000 inhabitants (http://fr.wikipedia.org/wiki/Abengourou). The Moyen Comoé region is the second most HIV-infected region in Côte d'Ivoire, with a prevalence rate of 5.8% (EIS, 2005).

According to Perrot (1981: 307), the Agni ethnic groups belong to the larger Akan ethnic groups, who have migrated from Ghana to Côte d'Ivoire around 1715.

[7] The regions where the PEPFAR-ANADER AIDS programme is implemented are: Bas-Sassandra (San Pedro), Lagunes (Dabou), Indenié (Abengourou), Zanzan (Tanda), Haut-Sassandra (Daloa), and Fromager (Gagnoa).

[8] The ANADER AIDS programmes have provided a medically equipped car with a mobile team of medical scientists, appointed to perform testing even in remote areas. The team collaborates with the local health center. The service is delivered on a from-time-to-time basis among the villages where the programme is implemented.

Originally, they came from a larger village called *Anyanya,* located near Enchi, the capital of Aowin in southwest Ghana. In *Anyanya*, 'they were living with the Baule, Akye (Attie), Sanwi, Morofwe'. These communities have all migrated from *Anyanya* to Côte d'Ivoire to escape from the intrusion and domination of the Denkyra and Ashanti, who were the most powerful ethnic groups among the Akan. The migration of the Baulé (or Baoulé) from Ghana to Côte d'Ivoire was led by queen Abla Pokou (Loukou, 1977), whereas the Agni migrated under the leadership of Ehuma Kablan and Ehora Amyao (Perrot, 1981: 311).

There are four smaller Agni ethnic groups in Côte d'Ivoire: Agni from Sanwi, located in southern Côte d'Ivoire, Agni from Morofwe (or Moronou), located in eastwestern Côte d'Ivoire, whereas Agni from Djuanblin (or Diabé) as well as Agni from Indenié (or Ndenye) are located in southeastern Côte d'Ivoire. Each community of these Agni groups is ruled by a king, with a kingdom composed of chiefs of cantons and chiefs of villages attached to the kingdom. According to the Agni political organization structure, a chief of a canton is the head of several chiefs of villages attached to the canton[9]. The Agni community living in the study area is the Indenié (or Ndenye) Agni group.

The larger Akan groups, including the Ashanti, Fanti, Baoulé, Attié, Nzema, and Agni, share the matrilineal kinship system, as the foundation of Akan culture. According to the Akan matrilineal system, only the nephews from the mother's side have the right to inherit from their maternal uncle, especially as far as the assets and royal positions are concerned. This implies that if a man dies, it is the son of his sister who will inherit from him, not his own son. Today, as we shall see in this study, maternal inheritance is often contested by the cousins of the matrilineal inheritor, who claim land rights through their father.

From the eighteenth century on, Abengourou has been the leading producer of cocoa and coffee in Côte d'Ivoire. For this reason, it was called *la boucle du café cacao*. During this period, the region was a prosperous one. This was visible through people's living standard, and the spread of luxury villas with air conditioning devices, even in some remote areas. Unfortunately, this period of prosperity coincided with the rise of HIV in the region. Some people believe that the district's prosperity in the past probably led to a way of life that increased people's exposure to HIV. It has been reported through interviews with local leaders that the prosperity led to increased polygamy in the district. Numerous men had two to four wives, and girlfriends as well. Besides, following this prosperous time, the region was hit by the disaster of a huge forest fire in 1997, which destroyed 4,000 acres of cash and food crops. From

[9] A canton is a central village to which several small villages are attached.

that period onwards, health, economic and ecological vulnerability in Abengourou have increased.

The study area is Y.F., one of the well-known sub-districts of Abengourou, and an important zone in terms of food provision. It is located at 240 kilometers from Abidjan, the country's economic capital and business center. The population is estimated at 29,151 inhabitants. The study area counts three primary schools, one public secondary school and one public health center, one post office, and a private timber company. It is served with a network of roads covering 240 kilometers. Most of the homes are supplied with electricity. Very few households have landline telephone connections, but many have mobile connections with Orange, MTN, and Moov, the three country-wide mobile service providers. There is as yet no internet connection in the study area.

The main economic activities are agricultural, but some small businesses offer an alternative source of livelihood. It is a multicultural village, with a majority of autochthonous Agni people, and migrants from other regions of Côte d'Ivoire and neighbouring countries such as Ghana, Burkina, Mali, Niger, and Guinea[10]. Y.F. is a cantonal village, i.e. a big and central village with ten villages attached to it. It was the first village of the Agni Indenié people. According to the history of the Agni, this village created the other Agni Indenié villages. The cantonal chief of Y.F. holds the longest reign record of chieftaincy in Côte d'Ivoire (sixty years). According to the Agni political organization structure, the chief of the village shares the power with the queen mother and the 'notables' who are entitled to assist the chief to rule the village. The 'notables' are the heads and chiefs of the fourteen big Agni families that form the village. Hence, there are fourteen notables at the royal court.

Agni community life and taboos

The life of the Agni communities cannot be pictured without the totems and taboos, observed in these people's everyday life world. According to Agni culture in the study area, Wednesday (*Mnan*) is not a day for farming, because according to the oral history, the totem, the spirit protector of the village, has settled in on a Wednesday. Therefore, the ancestors imposed on the people to honor this spirit of the village every Wednesday, by stopping farming that day. In addition, during the third Monday of the moon and the third Friday of the moon, women are not allowed to farm. If they do, the village may have to face the fury of the spirits. From local people's perspective, disrespect of cultural norms could be one of the reasons for the

[10] Among these migrants, Ghanese and other local ethnic groups who belong to the Akan group (Baoulé, Attié, Abbey ...) share the same cultural roots and kinship system with the native Agni, as they all originally came from Ghana. In this study, when I talk about patrilinearity as the kinship system of migrants, I consider the bulk of the migrants in the village who came from Burkina Faso, Mali, Guinea, Niger and Nigeria.

decline of the biodiversity of forests and rivers. For them, the environment belongs not only to humans, but also to spiritual powers, who are the main owners. In that sense, some days, which are called 'bad days', are culturally known to be reserved to the ghosts. During those days, people are not allowed to interfere or to interact with their natural environment.

Before starting any agricultural activities, Agni have to make a sacred area in the field, to honour the ghosts of the forest[11]. This aims not only at protecting them while performing their agricultural work in the forest, but also at improving and enhancing their agricultural production. For this reason, they annually have to perform rituals, by sacrificing a sheep or chicken to the ghost, i.e. the protector of the forest. When menstruating, women are not allowed to go near the seat of the chief in the royal court; if they do, they have to perform a ritual of purification, with the sacrifice of a chicken. Families can have a specific totem, which is mostly a food totem. For instance, some Agni families as well as the royal family venerate the *Tano* river. The members of these families are not allowed to eat goat or drink *bandji*, which is a local alcoholic drink made from the palm tree.

Agni women's 'power': the queen-mother and women's empowerment

In the matrilineal Agni communities, the chief rules the village with his mother, the queen-mother. In case of her death, she is succeeded by her eldest daughter, i.e. the sister of the chief. According to Agni political organization and hierarchy, the decision-making is shared between the chief and the queen-mother, who can rule the village when the chief is absent. The queen-mother incarnates socio-political authority. In that sense, one can state that women have organizational 'power' in Agni culture. Women's 'power' in the Agni matrilineal system is visible through their roles in production, reproduction, purification, and decision-making in the traditional political arena. Kings and chiefs are chosen exclusively by women. They are the ones who should decide and nominate kings and chiefs to rule the entire Agni community. In the study area, the traditional political authority is exerted by the queen-mother and her brother, the chief. Decision-making is shared and reinforced by the queen-mother, who often intervenes in private problems within households as well. From in-depth interviews, it emerged that she does her best, as much as she can, to resolve women's private problems. What is more, she often seems to be more powerful than the chief in these matters:

> *'There was a case of divorce in Pagnan, a village situated at 2 kilometers from here. The chief sent a group of people to ask for forgiveness, it did not*

[11] Even though the rituals to worship the ghost forest protector are still performed by people, the belief is not so strong as it was in the past, because religions such as Catholicism, Islam and Protestantism oppose such beliefs and practices.

> *work out! Some other initiatives undertaken to renounce the divorce have failed. But when the queen-mother herself went to ask for forgiveness, it worked out. This testifies to the importance of the queen-mother as part of traditional political authority, and her influence in conflicts management within and outside of households. Actually, Agni people are respectful of the traditional institutions. Their respect for the queen-mother confirms her "power", in Agni culture' [Kasi, Y.F, 15-02-08].*

In Agni culture, women have a role of purification and protection, performed through *moumoumin*, a ritual dance. The *moumoumin* dance is exclusively done by women, before any cultural events, with the aim to clean, purify and protect the event from bad spirits. During a crisis, such as war, women have to perform this dance to support and protect the warriors and the village. During the war in Côte d'Ivoire in 2002, women in the study area performed the *moumoumin* dance at the demand of the queen-mother. Women met at the royal court, wearing the traditional make up made of *kaolin* (granite) on their bodies, dressed in a piece of white cloth, and performed the dance everywhere in the village. While performing this dance, some women stripped off their clothes. For this reason, men in the village are informed in advance, and have to abandon the village during the performance. The *moumoumin* dance is relative to each situation and event. For instance, the *moumoumin* dance and songs that were performed during the war that Côte d'Ivoire was facing, were intended to protect the Agni who had moved out to fight, as well as those who had stayed in the village. Women took bananas, some piece of firewood, old pounder sticks, eggs, and ash, which they put at the main entrance of the village. For the women, the goal was achieved, since the out-migrated people came back safely and the village was protected from the war. It has also been reported that, whenever there is drought and women perform this dance at the demand of the king, it will rain. From an eco-feminist theory perspective, this shows women's connectedness to nature and their importance to respond to biodiversity degradation problems. In short, in Agni culture, women do not only play a role in reproduction, production and political decision-making, but also act as protectors of their people and nature.

However, in spite of their importance in Agni culture, as expressed through their decision-making power in the traditional political arena and their protective power in the cultural and ecological context, in everyday life women experience the weight of gender ideology within their households and have to play down their power. This is illustrated by the traditional Agni houses, which consisted of separated spaces for men and women. According to this design, women do not share the same space with men, regardless of their 'supposed power' derived from the matrilineal system. They meet only when they have to provide food in the men's space. Although this spatial separation does not exist anymore, the impact is still visible through the gender ideology in the Agni community. The man is the head and the decision-maker in Agni households. Household assets are controlled by men. FGDs with Agni men and Agni women show that, in Agni households, women cannot claim equality, autonomy and power in intra-households relations.

This area was chosen as a study site for various reasons. The village has AIDS-affected households from different cultures, i.e. Agni households and migrant households; it is an important zone in terms of food provision, faced with decline in food production during the last ten years.

2.5.2 Migrants among the Agni: the integration issue

Internal and external migration in Côte d'Ivoire is a historically well-known phenomenon. Socially, culturally, economically and politically, it has become very important to everyday life in most villages in Côte d'Ivoire, since they include many migrant Ivorians and migrants from neighbouring countries (Dureau, 1987; Fieloux, 1980; DSRP, 2009). In the study area, apart from migrants from neighbouring countries, communities such as the Baoulé, Ebrié, Bété, Dida, Sénoufos, Tagbana, Odienneka, who originate from the north, east, centre, south and west of Côte d'Ivoire, co-reside with the autochthonous Agni. However, kinship and religion create affinities and connections beyond the boundaries of ethnic groups. The Ivorian migrants who belong to the larger ethnic group of the Akan, have a matrilineal kinship system, and they usually have Roman Catholic and Protestant backgrounds. Most of the migrants from neighbouring countries have a patrilineal kinship system and are Muslim.

Although from a local leaders' perspective, there is integration between the autochtonous Agni and the migrants' communities, through participant observation, I could observe that they are not integrated geographically. They live separately, that is, not in the same streets. Agni households live together in one street called *Agnibougou*, i.e. the street of the Agni, whereas migrants' households live in another street called *Dioulabougou*, i.e. the street of the Dioula, or *Sohnsoribougou*, which is the local name of a township. I realized through interviews with Agni and local migrant leaders that communities are not willing to share the same geographic space with others for religious reasons, and feel more comfortable to live with their own group.[12]

During the FGDs with migrant men and women, I could see how they appreciate the leadership of the chief of the village. For most of them, although they do not have access to land, this village is one of the rare villages in Côte d'Ivoire where migrants have felt secure during the civil war. To them, this was possible because of the leadership of the chief and his commitment to protect them from the aftermath of the civil war. One of them remembers:

[12] In the study area, some households are mixed households, i.e. they are composed of a couple of both Agni and migrant origin. Members of such mixed households practice a religion of their choice.

'Policemen are not coming anymore to this village to ask for residence permits. One day they came, and went to the chief to tell him that they wanted to check the residence permits of migrants. The chief replied that he did not know any migrant in this village, since people who live in his village are his children and grand-children. From that day on, we never saw another policeman in this village asking to check our residence permits. In addition, by the time the civil war had become very hard, while some migrants were being killed in some villages, here, the chief set up a committee that was doing a routine to make sure that we were secure. The committee was meeting each Wednesday with the migrants' communities, to make sure that we were fine. Actually, we challenge any migrant elsewhere in this country, to show us that they have a chief whose leadership goes beyond the leadership of the chief of this village. So, any time we are requested by the chief to perform agricultural work for him in his fields, we always do so, and the motivation to perform the work is always high among the migrants. We still pray for him, for a healthy and long life' [Ibrah, Y.F, 11-03-08].

2.5.3 AIDS awareness and intervention programmes

In FGDs with Agni men and women, migrant men and women, youths, as well as interviews with local leaders and key informants, people have provided insights about the origin and spread of AIDS in the study area. For most of them, multiple factors are associated with the spread of AIDS in the area. They acknowledge that migration motivates people to engage in sex with new partners, increasing their exposure to HIV. It is also common that when people are ill in urban areas, some are inclined to come back to their rural home to look for traditional medicine, and therefore may contribute to an increased spreading of the disease among people in the village. Among the factors associated with the spread of AIDS in the area, people emphasized the decline in moral values. From their perspective, girls are more and more exposed to prostitution for cash, as they want to get a good life, like those living in big towns.

People agree that there is awareness about the AIDS epidemic and its impacts on people's livelihoods. They believe that this awareness is visible through the condoms thrown away everywhere on the streets, meaning that people now go for protected and safe sex. Unlike what was happening in the past, some people are now increasingly willing to go for the public testing organized by the PEPFAR-ANADER AIDS programme. At the village level, this is seen as a sign of AIDS awareness among the local people.

In the village, there are two AIDS mitigation programmes. The EGPAF programme is implemented by the local health center. The PEPFAR-ANADER programme is implemented by the ANADER mobile medical team, but works in collaboration with the health center. Both PEPFAR-ANADER and EGPAF AIDS programmes are implemented with the involvement and support of the village committee of the response to AIDS, composed of local leaders and local people.

2.5.4 Environmental degradation in the study area

Like in many areas in Côte d'Ivoire, the study area suffers from a lack of good soils that is affecting agricultural production. According to local people, this ecological vulnerability is also due to logging activities in the area. As the regional director of the national forestry agency[13] stated: 'Agricultural activities are the main reason of forest degradation in this area'. He accuses farmers who oppose the national forest protection programme, so they can make use of the *forêts classées*, i.e. the reserved forests, for farming. Consequently, in the district, from 1994 to 1995, 8,000 acres of forest were depleted by agricultural activities. Today, the forestry agency is dealing with reforestation activities, to stop forest degradation. In the course of the implementation of the forest stabilization and reforestation programme, through sensitization programmes, pressure was put on farmers to stop using the reserved forests for farming. 'Actually, agricultural activities have now been reduced in these spaces. So, we can state that we stabilized the degradation and are now dealing with reforestation', said the director, when interviewed on the subject.

In 1997, the Minister of Environment and the National Forestry Agency got the approval from the national government to destroy all kind of food crops in the reserved forests. For the director, this helped to stabilize the forests' degradation. In Côte d'Ivoire, SODEFOR and OIPR are appointed for forest protection. In the Moyen Comoé region, SODEFOR is managing 16 reserved forests, with a total estimated acreage of 34,500. According to the director, 30% of the cocoa produced in Côte d'Ivoire is grown in reserved forests. If they destroy these cultivations, he said, the economic impacts will be very strong. Therefore, as an alternative, they have decided to plant trees in cocoa and coffee plantations, so that in the future they will get the forest back. While asking to farmers to give up farming in the reserved forests, the director acknowledged the need to help farmers develop alternative livelihoods, based on non-forest resources. Approaching the problem from a gender perspective, he stated: 'we signed some contracts with women to cultivate the *tiama* plants that we use for reforestation. We provide them with seeds and afterwards we buy the plants from them. We also stimulate the cultivation of *Glicedia*, well known to minimize the impacts of forest degradation on soil fertility'.

[13] The National Forestry Agency (SODEFOR) is represented at the national level. OIPR is the National Office of Parks and Reserves. While SODEFOR is engaged in forest protection as its main activity, the OIPR is more concerned with the protection of animal species in danger of extinction.

Chapter 3.
Conceptual framework

This chapter presents and discusses the main concepts used as theoretical framework for the research. In the review, the concept of sustainable livelihood is discussed, followed by a discussion of the concept of household. Furthermore, the gendered impacts of AIDS and their implications for food security are discussed, as well as the concepts of kinship, vulnerability, resilience, food security, and environmental degradation, presented as part of the vulnerability context that impacts on households' food and livelihood security. The study design is then summarized in the theoretical framework. It has three parts. The first part concerns the factors influencing or determining women's vulnerability, the second is about the consequences of the implications for vulnerability (labour shortage, food insecurity), while the third part comprises women's agency in coping with their vulnerability.

3.1 Livelihood

The sustainable livelihood approach is well documented. According to the simple dictionary definition, livelihood, originating from the word 'to live', is a 'means of living'. The concept of livelihood in the social sciences has a more comprehensive meaning, and is widely used in poverty and rural development literature. According to Chambers and Conway (1992: 6), livelihood 'comprises the capabilities, assets (stores, resources, claims and access) and activities required for a means of living'. In the often cited definition by Ellis (2000: 10), livelihood 'comprises the assets (natural, physical, human, financial and social capital), the activities and the access to these (mediated by institutions and social relations) that together determine the living gained by the individual or household'. Livelihood security is linked to food security, because livelihood security means security in the provision of basic human needs such as food, clothing, shelter, education and health. Therefore, household food security is seen as a part of livelihood security. Frankenberger and McCaston (2001) relate livelihood security to food and nutrition security. Hence, if a household is livelihood secure, it is food secure. However the reverse is not always true: a food-secure household is not necessarily secure in terms of livelihood. In this study, we will add yet another aspect to our understanding of livelihood, because we feel that cultural values and practices also need to be addressed. According to Long (2001), the normative and cultural dimension of livelihoods leads to the exploration of the issues of lifestyles and the factors that underpin them. In that sense, the livelihood definition is not solely about economic resources but also about culture, which include values, identity, skills, perceptions, symbolic forms and organizational strategies. The literature shows a great variety of definitions of livelihood security and sustainable livelihood(s). Redclift (1990: 85) states that security is 'ownership of or access to resources and income earning activities, including reserves and assets to offset risks, ease shocks and meet contingencies', whereas sustainability refers to 'the maintenance or enhancement of resource productivity on a long-term basis'. Dietz (2000) believes that to be sustainable, a livelihood does not rely solely on one asset such as land, and livelihood improvement can be seen to rely on a variety of other options. Campilan (1998: ix) uses the definition of the World Commission on Environment and Development (WCED) for a livelihood as 'adequate reserves and supplies of food and cash to meet basic needs'. In addition, he found that 'sustainable livelihoods can be assured through: sustainable employment and adequate remuneration; engagement in productive activities which are ecologically sustainable and economically sound; and ownership of or access to resources and their management, within their capacities to recover'. In the definition given by Huq (2000: 117), 'livelihoods encompass income, both in cash and in kind, as well as social institutions relating to kinship, family, neighbourhood and village, women's groups and property rights'.

According to Niehof and Price (2001), in the definitions such as those of the WCED and Chambers, the distinction between the dimensions of process, activities, and outcomes is not clear. They suggest that livelihood generation takes place in the context of a multidimensional, dynamic system, called the livelihood system, that has the following components (Niehof and Price, 2001: 8-10; Niehof, 2004: 322):

- inputs: resources and assets;
- output: livelihood;
- purpose: livelihood adequacy to meet basic needs (Chambers, 1989);
- activities: livelihood generation and the composition of the livelihood portfolio;
- agency: efforts of households and individuals to achieve livelihood adequacy;
- quality: degree of vulnerability (or sustainability) of the livelihood produced;
- environment: context within which the livelihood system functions and interfaces with other systems and institutions;
- locus: the household as the locus of the livelihood generation.

Livelihoods can be assessed with regard to their degree of vulnerability. In other words, contexts in which households' livelihoods strategies are developed, involve a number of threats and shocks that render them more or less vulnerable. 'These threats can include periodic droughts, floods, pest infestations, crop and livestock shocks, economic shocks, conflict and civil unrest, as well as the illness and death of household members. Households are viewed as being sustainable if they can adjust to threats without compromising their ability to survive shocks to their livelihoods' (Stokes, 2003: 2). Brons *et al.* (2007) have pointed out the importance of including cultural variables in the livelihood framework. From their perspective, some cultural landscapes can potentially increase women's vulnerability. That may explain, for instance, women's greater vulnerability to AIDS impacts, compared to their male counterparts (see Chapter 5).

To overcome a situation of shock or stress, households mostly use assets such as savings, livestock, or social capital. Rural people's livelihood portfolios refer to bundles of carried out activities (Niehof and Price, 2001). A rural livelihood portfolio generates livelihood security by using assets and resources as inputs. Livelihood strategies include the use of all the assets and resources that rural people have access to, such as human resources (labour, education, skills, experience and health), natural resources (land, water), physical resources (houses, roads, irrigation canals, machines, tools, etc.), financial resources (savings, credit), social capital (networking, labour-sharing), as well as cultural resources (values, lifestyles, identity, skills, perceptions, symbolic forms, and organizational strategies) (Long, 2001; Brons *et al.*, 2007). The dynamics and the level of household livelihood security are a function of the availability and the access to these assets and resources. Therefore, livelihood security or sustainability is based on access to resources (Berry, 1989; Blaikie, 1989), the composition of the livelihood portfolio (Niehof and Price, 2001), asset vulnerability (Moser, 1998), entitlements (Sen, 1981), livelihood diversity (Ellis, 2000), and cultural dimensions of livelihood (Long, 2001; Brons *et al.*, 2007).

While the livelihood framework aims at assessing the causes and solutions of rural poverty, the linkages between gender, rural livelihood and food security are still missing in many studies. Despite the fact that women provide the bulk of the food and agricultural production, they are disadvantaged with regard to entitlements, the ownership of assets, and access to and control over resources, such as land and credit that are essential for agricultural production and livelihood generation. It is noteworthy that the lack of access to assets and resources and the poverty resulting from it render women more vulnerable to HIV and AIDS, as it increases their exposure to prostitution and transactional sex in order to feed their families. Women's vulnerability to AIDS impacts seriously threatens rural households' livelihood and food security. In the context of AIDS, declines in human capital through the illness or death of one or more household members affect the other livelihood assets (Tapouzis and Du Guerny, 1999; Müller, 2005a,b; Nombo, 2007). With the loss of experienced agricultural workers, human capital declines are multidimensional, resulting in labour shortages and decreasing productivity both on and off the farm, which consequently leads to declines of household income. Increasing household expenditures on medical care and funeral ceremonies result in a decline in household savings and the loss of assets through the sale of both productive and non-productive assets. The decrease in production, leading to a decrease in income, leads to a decrease in food consumption and nutrition status.

In short, although women remain the key actor of food production, food provision and food security, these issues are still insufficiently incorporated into livelihood and gender theory. This is strongly affecting the relevance of AIDS mitigation policies in sub-Saharan African countries.

In this research, a livelihood perspective is adopted to understand how AIDS affects both the food and livelihood security of both autochthonous Agni and migrant households, and how women cope with vulnerability to uphold their livelihood generation. Where many food security studies mainly provide institutional and demographic data, this study purposely integrates demographic data with ethnographic data on cultural differentiation, to show how they influence women's vulnerability in terms of sexuality, access to land and food crops, and the availability of agricultural labour. Drawing on an actor-oriented approach, this research goes beyond the sustainable livelihood approach that has failed to address the value contestations that shape the ways in which people interact with each other and cope with vulnerability in everyday life. In short, I concur with Arce (2003: 203) that the sustainable livelihood approach focuses more on economic features than on the 'social value and actors' understandings of their own reality'. People may challenge sustainable development agendas whenever these do not fit and respond to their 'reality'. This study adopts a livelihood perspective that draws on 'a more dynamic approach to the understanding of local livelihoods, which stresses the interplay and mutual determination of the contestations of values and relationships and which recognizes the central place played by human action and identity rather than capital' (Arce, 2003: 204).

3.2 The household

In many livelihood studies, the household is regarded as the micro-level unit of analysis, i.e. a basic unit of human social organization, representing the arena of resource sharing and exchange that people perform in everyday life (Clay and Schwartzweller, 1991; Bruce and Lloyd, 1995). This makes the household the locus of livelihood generation (Niehof, 2004), a place of confrontation and reproduction of social norms and values, power and privileges (Narayan *et al.*, 1999). Ellis (2000) emphasizes co-residence, eating together, and coordinated decision making on resource allocation. The meaning of household is relative to the form, the function, as well as the socio-cultural landscapes (Netting *et al.*, 1984; Guyer and Peters, 1987). In the African context, the household is seen as dynamic and flexible. It refers to a wider and strong network of family and kin, with boundaries that are not only fluid, but also vary in relation to social and economic differences (Nguthi, 2007; Karuhanga, 2008). Households are family-based, but may include non-family members like friends or servants, who are not biologically related (Rudie, 1995, Pennartz and Niehof, 1999). Even though household refers to a common residence, household members may also include those who are not in residence, but contribute to the household income with remittances and other necessities and services (Netting and Wilk, 1984). In short, households change their composition and boundaries over time, among others through migration and polygamy.

The concept of household as an economic unit has been criticized by feminists, for being gender-blind (Kabeer, 1991). For them, gender inequalities determine the intra-household relations, which may lead to conflicts as well as cooperation, and may result in unequal power between its members. Therefore, livelihoods and individual well-being are seen as the outcomes of negotiations and bargaining between households members (Sen, 1990; Moser, 1993; Karuhanga, 2008). The moral dimension of the household economy is highlighted by Cheal (1989), and Pennartz and Niehof (1999). According to them, there is a moral content to domestic arrangements, applied to sharing resources and caring for dependants in the household. To this extent, we can appreciate household's utility to respond to care-related demands in a context of HIV and AIDS (Niehof, 2004b; Nombo, 2007).

In the study area, even though the autochthonous Agni recognize matrilineal rights to inherit land, they take up patrilocal residence, as do the patrilineal migrants. In this study, I consider the household as the primary unit of analysis, defined as a group of individuals who are related by either blood or marriage, including unrelated persons like friends and temporary dwellers, who share a common residence, meals, provide things for each other, and acknowledge the same household head. This includes members of the household who may not be physically present sometimes, and excludes the members that have permanently migrated, even if they provide income or support as part of the household's resources.

3.3 Gender and AIDS

Gender is a critical conceptual category. It came up in the 1990s and has been related to food security in many studies. Gender is socially constructed femaleness and maleness, varying across cultures. The concept refers to the psychological, social and cultural differences between men and women. According to Kabeer (1995), social norms are gendered as well. In fact, regardless of the key role that women are playing in the socio-economic development process, their contribution generally remains invisible, due to social, economic, political and cultural gender inequalities. The socio-cultural perceptions of women's contribution to the household affect their bargaining power at the household level (Sen, 1990). Women's position in society tends to be associated with marginalization and poverty, as their status is known to be lower than that of their male counterparts. This is due to their lesser access to education, entitlements, resources and assets, as well as their burden of productive and reproductive responsibilities, leading to a feminization of poverty (Moser, 1993; Firebaugh, 1994; Chant, 1997).

In spite of their relative poverty due to gender inequalities, it is noteworthy that women spend cash resources on household basic needs more than men do, which is important for child nutrition. Gender and household food security are related, since household food and nutrition security remain women's concern. The FAO (2004) acknowledges that in sub-Saharan Africa, 80% of the economically active women work in the agricultural sector. Therefore, they feed the region. It is also noteworthy that in households headed by a woman with access to resources and assets, food security, education and health are better guaranteed. Therefore, gender equality is seen to be crucial for household food and nutrition security (Quisumbing *et al.*, 1998; Smith *et al.*, 2003; Müller, 2005a,b).

Household as the unit of analysis plays a crucial role in gender analysis. Intra-household relations determine the access to and control over resources, based on a sexual division of labour. Men's labour is known to be centered on production for the market economy, whereas women's labour is utilized in both reproduction and production in the unpaid economy. Given the gender division of labour, as a consequence, women's contribution to the process of production remains rather invisible (Moser, 1993).

The role of gender in both household decision making and its livelihood strategies is important, which is why an assessment of household livelihood strategies requires a gender-based analysis of its members' roles in various activities, their access and control of resources, and their role in decision making. Such an analysis also allows a better understanding of the feminization of poverty and of women's needs. According to Moser (1993), gender needs are practical and strategic. Practical gender needs are the needs that women identify in their socially accepted roles and subordinate position, with regard to the gender division of labour. These needs relate to women's experience in their everyday life world, mostly to water provision, means of transportation, health care, and employment. Strategic gender needs are needs perceived as a result of women's subordinated position in society, relating to gender norms and roles, power and control. Strategic gender needs refer to women having access and control over resources and productive assets, education, credit, organizations, social networks, legal rights, and a real integration in the political system. 'Strategic gender needs vary according to cultural contexts and include needs such as ownership and inheritance rights, protection from domestic violence, equal wages and opportunities, and women's reproductive rights' (Wiegers, 2008: 31-32).

Gender has been identified to play a key role in the HIV epidemic and AIDS impacts in sub-Saharan Africa. Susceptibility to HIV is not gender-neutral. Cultural landscapes make men and women more or less susceptible to contracting HIV (Müller, 2005a; Wiegers, 2008; Niehof and Price, 2008). In most of the sub-Saharan African countries, for men to have multiple partners fit in with values of manhood, which increases a households' exposure to the risk of HIV. In most sub-Saharan African countries, the cultural rationale behind women's reproductive role, i.e. motherhood, represents an obstruction to condom use. Fecundity remains highly valued by cultures in these countries, and characterizes women's social position. In such a context, condoms are seen as a potential danger to cultural survival. This deepens women's stigmatization as well, since to stop childbearing means renouncing their reproductive power, i.e. their 'key power' in society.

The autochthonous Agni women's power lies in their reproductive role. Agni culture and the entire community rely on women to sustain the descent line and the chieftaincy by the provision of children. This study shows that while Agni women see bearing children as their duty *vis-à-vis* their culture, some may have several partners before getting married. Although migrant women have to keep their virginity up to marriage, nowadays, some of them engage in transactional sex, sometimes with several partners, before getting married. In this way, women may expose themselves and their partners to the risk of an HIV infection.

In many African societies, women suffer from unequal sexual relationships between women and men. Male sexual desire and men's control over sexuality are perceived as having priority (Gupta, 2002a; Wiegers, 2008). In this study, we found that married migrant women are supposed to be sexually passive and submissive *vis-à-vis* their husbands. Results show that, most of time, they do not have control over their sexuality or the sexual behaviour of their husbands. They do not dare ask for safe sex with condoms, as they have to respect the cultural norms. Therefore, this study does not corroborate the findings of Orubuloye *et al.* (1993), who found that, in West Africa, women have more control over their sexuality, as they have more economic power and may rely on their own family for financial support to cope with hardship in their household.

However, these researchers overlooked the micro-level, specific context that reveals how social dynamics shape sexuality in gender relations, and how this varies between cultures within a country and beyond. Looking at differences between the Agni and migrant cultures in Côte d'Ivoire, this study shows that Agni women have control over their sexuality before getting married but lose it once they get married, as their culture does not allow them to refuse the sex desired by their husbands. Migrant women have to preserve their virginity. Once married, they do not have control over their own sexuality; the husband's sexual desire comes first. Even in an AIDS context, they have to oblige, otherwise they will not escape stigmatization in their community. In short, one can see that, in West Africa, women's control over their sexuality is relative, i.e. context-specific.

In conclusion, the causes of women's susceptibility to HIV and vulnerability to AIDS impacts are biological, rooted in sexual behaviour, cultural norms, social attitudes, and livelihood vulnerability. Therefore, women's vulnerability in a context of HIV and AIDS are gendered, and culture-specific. Thus, I concur with Machera (2004), that the high rate of women's HIV infection is related to women's lack of *bedroom power.* In short, 'If more women have the power to "say no" to unwanted and unsafe sex, the HIV infection rate would dramatically decline in Africa' (Machera, 2004: 167). The graver the HIV epidemic gets, the more reasons there are to focus discussions on gendered power relations in the context of sexuality (Arnfred, 2004). It can be concluded that 'prevention of the spread of HIV can only be achieved through greater de facto gender equality' (Burns, 2002: 6-7).

3.4 Matrilineal and patrilineal kinship

Kinship has been defined and interpreted differently by different scholars. According to Niehof (1999; 2003a), kinship refers to relationships based on birth, marriage and to relationships framed in this idiom. Social scientists see kinship as much more than biologically determined relationships, as it involves the wider social and cultural institutions in which social interaction is embedded, like in Bourdieu's *habitus*. The role and functioning of kinship is culturally specific in the ways it organizes the functioning of households, people's livelihood strategies, and the social structure of local communities (Pennartz and Niehof, 1999). So, kinship studies and analysis give insight into people's access to assets and resources, rights and entitlements, caring-related emotional support, agricultural production, and migration. Kinship does not only play a role in so-called traditional societies, but it remains important in most present-day African societies, as well as in people's social and cultural interactions. In our research area, two kinship systems exist side by side, that of the matrilineal Agni and other Akan migrants from within Côte d'Ivoire and Ghana, and that of the mostly patrilineal migrants from the North.

Kinship can be traced through descent in three different ways (Fox, 1967; Niehof, 2003a). In a patrilineal kinship system, it is traced through the father's line. Descent is traced through the mother's line in a matrilineal kinship system. A bilateral kinship system traces descent through both the father and the mother. In a patrilineal kinship system, children belong to the patrilineage of their father, and name, land ownership and entitlements are transferred from fathers to sons. In a matrilineal system, children belong to the matrilineage of their mother, and land is transferred from mother to daughter, or from father's sister to brother's son.

The matrilineal system of the Akan ethnic group has been discussed by several scholars. According to Kofi (1990), the Akan language is full of statements about the matrilineal relationships. *Abusua*, or family, refers to the matriclan and not to the patriclan, because 'Akan children inherit their soul or spirit (*ntoro*) from their father and their flesh and blood (*mogya*) from their mother. Since the blood and flesh are the source of existence, children are believed to be related to their mother, hence a member of their mother's family' (Kofi, 1990: 5). So, in the Akan matrilineal system, the corporate clan is dominant over the conjugal family. Therefore, with regard to the Akan matrilineal system of inheritance, 'a son does not inherit from his father. Rather, fathers are expected to set up their male children in life through training, giving of gifts, and helping sons to acquire their first wife ... since it is matrilineal, a daughter or son can inherit the self-acquired property of the mother, but a son cannot inherit that of his father' (Kofi, 1990: 7). The matrilineal system has led to a 'nephew inheritance' system among the Akan.

Among the West African Akan people, like the Ashanti, Baulé, and Agni, especially among royal lineages, land is inherited through the matrilineal line, from a man's sister to his son. In other words, according to this rule, his own son does not inherit from him. In short, 'the matrilineal inheritance is part of a complex network of social structure which gives not only security but also an identity to the Akan. The matriclan is the basis of social organisation. One's right to property, such as land, status, and position in the traditional structure is based on one's membership in a matriclan' (Kofi, 1990: 12).

This study confirms that among the matrilineal Agni, indeed only nephews inherit the royal assets, the chief's positions and his political power from their maternal uncles. According to the Agni matrilineal system, land is inherited through the mother's line as well. With regard to the social-political hierarchy embedded in the Agni matrilineal system, the queen-mother is entitled to rule together with her son, and if she is not there, it is the sister of the chief (or king) who takes her position. This is the case, for example, in Y.F., where the sister of the chief rules together with him. At times she is even more powerful, because she inherited her position from the 'queen-mother' in the Agni community. In the political order of Y.F., the sister of the chief is called the 'queen-mother'.

There is a complex relationship between descent and residence in matrilineal systems, in the sense that geo-political alliances are made between men who are from the same matriline, but who are living in different villages. The successor to a chief or ruler seldom lives in the same village, and almost always comes from outside (see Van den Breemer, 1984: 157-158). In matrilineal societies, upon marriage, the husband moves to the home of his wife. Residence is usually matrilocal or uxorilocal, when a couple lives with the family of the wife, or avunculocal when they live with the mother's brother or uncle (Niehof, 2003a). According to Duchesne (1994), although the Akan kinship system is matrilineal, residence is most of time virilocal. In the case of the Agni of Y.F., we see a pattern of a combination of matrilineal descent and patrilocal residence, the couple living with the family of the man. Consequently, we could regularly observe tension between the husband's sons and their matrilineal cousins about land inheritance. Therefore, there is a conflict of authority inherent in matrilineal kinship with regard to the system of inheritance. Some people challenge the customary system of intestate succession, leaving their property to their own children instead of the culturally legitimate inheritor, the nephew. This practice has brought a kind of shift from matrilineal to patrilineal descent (Richards, 1950; Cardinal, 1931; Douglas, 1969; Kofi, 1990). The conflict inherent to the matrilineal inheritance system has led to the promulgation of laws in Ghana (Kofi, 1990). In Côte d'Ivoire, patrilineal inheritance is now accepted by the national Law [N°64-

379] of 7 October 1964, which creates a legally pluralistic situation when matrilineal descendants and patrilineal claimants contest the ownership of the same land.

Accordingly, in a context of inter-ethnic marriages between the matrilineal Akan and patrilineal groups, 'if an Akan woman marries a non-Akan man, the children have dual inheritance. They can inherit through both parents. On the other hand, if the marriage is between an Akan man and a non-Akan woman, the children have no automatic inheritance in either the matriclan or the patriclan. Wives and children in the latter union, if contracted under Customary Marriage Law, may be left uncared for when one of the spouses dies intestate' (Kofi, 1990: 9-10).

This study confirms that in patrilineal societies, like in the case of the migrant women, women have no rights to land. In their own family, only brothers can inherit land from the father. Even though they provide the bulk of the work on their husband's family's land, they have no rights to it. They are not entitled to inherit land from their husbands. In our study area, migrant women do not have access to land, unless they rent it for crop production. But many of them cannot afford the cost. For this reason, they usually work on the plots of their husbands to provide food for the household.

Kinship networks within and outside the village evidently provide social security to their members. Therefore, kinship has a moral dimension and involves obligations that become visible in caring, particularly for the elderly, and not only within households, but also beyond households and even villages. According to Niehof (2003a: 9), 'in rural areas in Africa, where there is a high prevalence of HIV/AIDS, the role of kin in caring for AIDS patients and their families is very essential. For women particularly, this is a heavy burden. When women have no land of their own, like in patrilineal societies, they often cannot cope in sustaining their household when their husband falls ill and dies'. Kinship obligations, like attending funerals, are another example of the burden of kinship in the context of AIDS.

3.5 Vulnerability

The concept of vulnerability has a key place in both the livelihood debate and disaster literature. Vulnerability is the opposite of security. In other words, when the livelihoods are insecure, they are said to be vulnerable. In the context of the literature on poverty, vulnerability is seen as a symptom of poverty (Wisner, 1993; Banerjee and Newman, 1994; Morduch, 1994). However, according to Devereux (2001), livelihood vulnerability or insecurity is rather a cause that contributes to poverty, than a symptom of poverty. In fact, if poor households are among the most vulnerable, rich households may also be vulnerable due not to poverty, but to certain shocks, such as the death or illness of the household head. Thus, vulnerability is not just a cause but also an outcome, and is part of a dynamic process (Glewwe and Hall, 1998; Moser, 1998; Bebbington, 1999).

Social scientists have put forward different definitions from different perspectives. Vulnerability is seen as an exposure to all kinds of risks, shocks and stress. For Chambers, vulnerability is not only an exposure to risks, shocks and stress, but also the inability to overcome these risks, shocks and stress. Vulnerability is then defined in terms of the risk for a household to fall into poverty in the future (Pritchett *et al.*, 2000; Christiaensen *et al.*, 2000). To Moser (1996), vulnerability is due to a lack of assets, whereas for Swift (1989), it is more about the management of assets. In Huq's (2000) perspective, the definition of vulnerability is focused on the correlation between income and access to resources, and on people's capability to protect themselves and recover from shocks.

According to some authors, there are different types of vulnerability from different perspectives. For Alwang *et al.* (2001), vulnerability is divided into several components of a risk chain: (a) the risk; (b) the responses; and (c) the outcome in terms of welfare loss. Sen believes that the classification of vulnerability is about risk and rights. He states that 'the first may be called the "risks-centric view", whereby vulnerability is typically defined as variability in the living standard caused by consumption or income shocks. The second perspective may be called the "rights-centric view", whereby vulnerability is caused by the lack of social and political rights' (Sen, 2002: 1). In Sen's opinion, there is a gender perspective to vulnerability. He states that gender vulnerability is a function to gender hierarchies and gender inequalities in the development process, with regard to the ways in which women experience marginalisation and discrimination, compared to their male counterparts. He also states that exclusion is the major cause of social vulnerability, due to the collapse of social ties, discrimination and marginalization experienced by minority groups, and the lack of protection for vulnerable persons, such as widows and the elderly.

Brons, *et al.* (2007) believe that the vulnerability of livelihoods in less-favoured areas lies with internal and external dimensions, and is linked to dynamics experienced at both the individual and collective level. 'These linkages between the external and internal dimensions of vulnerability and responses at the individual, aggregate and collective level should be studied to understand and mitigate current trends of increasing vulnerability of livelihoods in less-favoured areas' (Brons *et al.*, 2007: 1). The distinction between an internal and external side to vulnerability is highlighted in livelihood studies (Chambers, 1990), and related to the distinction between micro-level strategies and macro-level outcomes (Krishna, 2004), and between the individual and the collective (Rudd, 2003). Internal and external vulnerability are assessed in relation to ecology, institutions, culture and economy. Less-favoured areas are areas where vulnerable livelihoods converge with vulnerable environments (Brons *et al.*, 2007). In this context, Blaikie *et al.* (1994: 9) define vulnerability as 'the characteristics of a person or group in terms of their capacity to anticipate, cope with, resist and cover from the impacts of natural hazard'. Ecosystems' degradation, such as forest degradation or unpredictable rainfall and high evapo-transpiration refer to external ecological and environmental vulnerability which seriously impacts livelihoods (Fraser, 2003; Dietz *et al.*, 2004). In Chambers' (1990) perspective, the external sides of vulnerability is visible through risks, shocks and stress that individual and household face in their life cycle, whereas the internal side is about the ability to overcome those without irreversible loss of assets. According to Brons *et al.* (2007: 3), 'the internal side of vulnerability is a result of entitlement failures and lack of access to certain capitals'. Thus, the vulnerability context in less-favoured areas refers to the biophysical environment and the societal context. In other words, less-favoured areas are characterized by poor natural resources (ecological vulnerability) and poor institutional infrastructures (societal vulnerability).

The livelihood framework distinguishes a number of capitals. 'These capitals include contextual features such as physical and institutional infrastructures and societal norms, as well as individual assets' (Brons *et al.*, 2007: 3). In many sub-Saharan African countries, AIDS is known to be part of the external vulnerability context of rural households (Blaikie *et al.*, 1994; Barnett *et al.*, 2002). In terms of internal vulnerability, Brons *et al.* (2007: 4) believe that 'the internal dimension of vulnerability has to do with the characteristics of individuals (age, sex, education, skills, health status, etc.) and households (gender of headship, income, asset ownership, dependency ratio, etc.) and other micro-networks'. According to Niehof (2004), liabilities can be seen as the opposite of assets. So, in a context of gender inequality to the disadvantage of women, for women, their gender becomes a liability.

In many livelihoods' studies, resource scarcity creates social processes that increase people's vulnerability. In addition, institutional dynamics depending on social organisation can increase people's vulnerability. For instance, unequal access to land due to local practices of lands management increases poverty. Such institutions and practices reinforce class and gender inequality. As such, 'overcoming poverty requires overcoming culturally-underpinned normative structures that keep people poor' (Brons *et al.*, 2007: 5). However, such culture-based structures are not always visible in current livelihood research (Fraser, 2003). Hence, there is a need to refocus the livelihood framework on access to resources and power relations that define entitlements, in which gender is a key issue (Negash, 2001; Niehof, 2004).

The local economy can impact and increase people's vulnerability in different ways (Haggblade *et al.*, 2002; Campbell *et al.*, 2002; Brons *et al.*, 2007). For example, a lack of market integration due to a lack of infrastructure, such as roads, increases people's vulnerability and poverty. Thus, technical and socio-economic responses are distinguished to overcome household livelihood vulnerability. Indeed, technological interventions such as the application of soil and water conservation techniques in resource conservation may improve a household's livelihood. Socio-economic responses are visible through income diversification, and in social networks used for social and cultural security. According to Brons *et al.* (2007: 7), 'responses that are strongly culturally underpinned often only receive marginal attention in livelihood studies, yet culture is important in providing or constraining options and shaping responses'. Cultural factors influence livelihood, AIDS and gender. Women who have no means of living (internal vulnerability) are exposed to AIDS risks (external vulnerability), as they likely engage in prostitution and transactional sex to cope with economic hardship.

Drawing on an actor-oriented approach, this study shows how Agni and migrant women experience internal vulnerability because of their limited access to education, assets and resources; and external vulnerability due to their limited access to credit and agricultural inputs, hired agricultural labour, marketing facilities, and health facilities. The study shows that while they are challenging their vulnerability, women appear quite resilient against these structural conditions of vulnerability, using their agency to create new practices, rules and resources to improve their position within the social arena. At the same time, however, they run the risk of increasing their exposure to HIV, jeopardizing the very objective of sustainable food security. In short, this study looks at the social and cultural dynamics that underpin women's experiences and responses to vulnerability.

3.6 Resilience

The concept of resilience has been widely discussed by scholars. According to Walker *et al.* (2004: 1), resilience is 'the capacity of a system to absorb disturbance and reorganize while undergoing change so as to still retain essentially the same function, structure, identify, and feedbacks'. In the literature, resilience captures a 'bounce-back-ability' (Davis, 1993). Scholars have explored the dimensions of the concept of resilience and its applicability and usefulness in the context of ill health and adversity (Loevinson and Gillespie, 2003; Wiegers, 2008; Niehof, 2008; Nombo and Niehof, 2008). In a context of high prevalence of HIV and AIDS and other adversity, resilience refers to people's ability to challenge the devastating impacts of AIDS, i.e. improve their health and regain a certain well-being, in areas such as their income, education, and food security (Loevinson and Gillespie, 2003; Niehof, 2008). The evidence in the literature shows that resilience is determined by internal strength and external factors that challenge it. Thus, Carpenter *et al.* (2001) argue that resilience has to be addressed as resilience *of* what and *to* what. However, with regard to the focus of this thesis, i.e. the resilience of Agni and migrant women in a context of ill health and adversity, I agree with Niehof (2008: 218) that 'it is more appropriate to rephrase the two sides of the coin of resilience of Carpenter *et al.* as resilience of *whom* to *what*'. However, 'the two dimensions of resilience raise a question about their interrelationships. How do they interface?

Does the kind and severity of the challenges and shocks people face, affect their strength or do people have a given amount of strength irrespective of the challenges they are confronted with? This leads to further questions, such as: why are some people stronger than others and what are their sources of strength? What causes some people to be able to bounce back even under the most adverse circumstances while others cannot?' (Niehof, 2008: 218). Davis (1993) believes that the sensitivity related to the shock experienced by a livelihood system determines its resilience, i.e. its 'bounce-back-ability'. According to Kessy and Obrist (2008), resilience is part of a multilayered process. To Niehof (2008: 218), 'people's agency can be positioned as a mediating factor in the interface between their sensitivity/vulnerability and the external vulnerability context, which – at the end of the day – together determine the outcome of the process in terms of resilience'. Niehof (2008: 218) sees agency as 'conscious and purposive behaviour of actors who consider the appropriateness and efficacy of their behaviour in a certain normative and institutional context'. In his paper on resilience, Van der Geest (2008) shows how culturally underpinned reciprocity between parents and children provides space for resilience, especially in the case of old-age dependency. In his view, old people and their children may use their agency for reciprocal moral and normative support, although reciprocal social relationships that play an important role in people's resilience need nurturing and investment to be effective and sustained. In Matinga's (2008) view, resilience as the ability to bounce back differs from the ability to endure hardship, which she refers to as 'hardiness' or 'toughness'.

Niehof points to the temporal dimension of resilience, although it is difficult to measure. This applies especially to resilience to AIDS impacts, 'because of the complex temporal and spatial dynamics of the epidemic' (Niehof, 2008: 222).

The question on how the layers that underpin resilience as a process are socially, culturally and institutionally embedded, including the question of how resilience is gendered, remain key questions from a social science perspective. This study shows the resilience of households and particularly of Agni and migrant women to the vulnerability of AIDS-induced morbidity and mortality. In this study, I refer to Davis's definition (1993) of resilience as the ability to bounce back, and the definition of Matinga (2008) as the ability to endure 'hardiness', that is embedded in cultural values. This study intends to contribute to understanding the subjective and qualitative dimensions of resilience, particularly how it is socially, culturally and institutionally embedded, and how resilience is gendered.

3.7 Human agency

Since the end of the 1970s, structural or institutional approaches are being contested by authors like Giddens and Long. They give more analytical space to individual and institutional actors who, to a large extent, construct their own lifeworld by actively using their agency to bring about change. This debate is often being referred to as the structure-agency debate. From this debate, theoretical and methodological efforts have emerged, for overcoming the dualisms of 'micro-macro', 'structure-action', 'individual-collective' (Knorr-Cetina and Cicourel, 1981). In this study, we will come across institutional frames and external constraints that women experience in shaping their livelihood and safeguarding food security. Vulnerabilities can be explained, for example, through quantitative data on morbidity and mortality. Yet, at the same time, our ethnographic study shows a great variety and differentiation in women's individual and group endeavours to challenge the structural demands of kinship and inheritance rules, or cultural norms about sexual behaviour. These differences cannot always be explained by a reference to abstract structures. The everyday action and valuation of actors are part of the structuration process (Giddens, 1984, see Section 3.7.1). Although cultural history or *habitus* (Bourdieu, 1977, see Section 3.7.2) may provide the justification for a present action, the actual differentiation of human action demands a more actor-oriented approach.

3.7.1 Giddens' structuration theory

In Giddens' structuration theory, it doesn't make sense to analyze action and structure separately, as structure is shaped by human agency, while actions do not have significance otherwise than through the background of structure. In Giddens' perspective, structure consists of rules and resources involving human action. While the rules constrain the actions, the resources make them possible. Giddens argues that systems are 'the structural properties of the social medium and the outcome of the practices they recursively organize' (Giddens, 1984: 25). Social structures are constituted by human agency, while at the same time actions are the medium and outcomes of social structures. In his structuration theory, Giddens overcomes the dualism between structure and agency by recognizing structure as a process. Thus he 'transforms' the structure as a deterministic 'societal frame' into a changeable 'process' (i.e. 'both medium and outcome of practices'), in contrast to functionalist theory, which did not create space for human agency (Otsuki, 2007: 33). Rules and resources are organized as the properties of social systems that consist of 'the situated activities of human agents reproduced across time and space' (Giddens, 1984: 25). In short, Giddens' structuration theory is a practice-based reinterpretation of society, where human agency is not opposed to structure, but a component of structure (Sewell, 1992).

3.7.2 Bourdieu's theory of practice

While Giddens reconceptualises the structure-agency dualism through his theory of structuration, Bourdieu conceptualizes *habitus* through a theory of practice. *Habitus* displays social structures, constituted by the dispositions of agents, shaped by their positions within the 'social field' (Bourdieu, 1977). In his conceptualization of *habitus*,

Bourdieu considers agent's dispositions and social positions as essential elements of the reproduction of social space. Agents' positions shape their dispositions. In lieu of rules and resources that shaped agents' actions in Giddens' structuration theory, Bourdieu indicates different capitals, including symbolic capital, that take into account the agents' construction of the social world, as they 'transcend objectivism and subjectivism' (Bourdieu, 1992: 126). So, in their contribution to the understanding of human agency in relation to structure, Giddens and Bourdieu portray agency as a social element, and structure as a 'process of ordering' (Long and Van der Ploeg, 1995). In their perspective, individuals are 'agents' who can exercise their agency to reconstitute structure, rules and norms. Agency is socially constructed. By acquiring agency, 'the individual' becomes the 'social actor'.

Drawing on Bourdieu's theory, we will see in Chapter 5 that matrilineal and patrilineal systems are objective social structures that are incorporated into the subjective experience and actions of Agni and migrant women. Hence, Agni and migrant women are both object and subject of their respective *habitus*, shaped by their respective cultures. The increasing importance of fulfilling their reproductive role and the related risks for Agni women; the submissive attitude to their husband and its related risks for migrant women; both are a set of acquired patterns of thoughts and behaviour, adopted by Agni and migrant women through their upbringing and education. Therefore, the concept of *habitus* lies at the individual level, as it is 'a system of acquired dispositions functioning on the practical level as categories of perception and assessment ... as well as being the organizing principles of action' (Bourdieu, 1990: 13).

Matrilineal and patrilineal kinship rules apply to the Agni and migrant groups, respectively. As institutions, they comprise the social structure of both Agni and migrant groups, i.e. their respective norms, values, social roles that make up their social systems, and social behaviour. Beyond the demographic figures, the matrilineal and patrilineal kinship systems also induce socio-cultural inequalities that likely increase women's vulnerability to AIDS and related impacts.

3.7.3 The actor-oriented approach

The actor-oriented approach was formulated to provide due space to and focus on human agency and its relation to social change in development theories and practices. The actor-oriented approach has compact theoretical foundations, based on critiques of development ideologies and practices that appeared to be 'structuralist' (Long and Long, 1992). It is embedded in a 'social constructionist' view of development that emphasizes social relations as 'constructed within development arenas and shaped by the interlocking of actors' projects, resources and multiple discourses' (Long and Van der Ploeg 1995: 66). The actor-oriented approach involves understanding power, action and meaning in social spaces as being embedded in the ways people respond to development interventions. It is based on the idea of 'social interfaces' that underpin people's daily interactions. Different actors have different powers, means and strategies, different standpoints and perspectives that shape and configure social life. Therefore, an actor-oriented approach illustrates the interactions between various actors, and the interfaces between actors and their institutions. It maps the existing situation, to identify the value contestations that underpin actors' everyday life world, and to provide a rationale for interventions (Arce, 2003; Long, 1989, 2001). Social analysis is based on ethnographic methods that allow the researcher to experience the everyday dynamics of the actors' social life.

In this study on women's agriculture-based livelihood and food security in south-eastern Côte d'Ivoire, it is evident that livelihood strategies and actions necessitate human resources shaped by knowledge, skills, physical power, strength, as well as good health. According to the DFID (2000), people need such resources to achieve their livelihood objectives. It is apparent that human resources are a major asset, as they shape livelihood outcomes. Human resources are turned into functions whilst they are used as inputs, in the course of livelihood generation. People may use a higher education level and/or other resources, for example, to earn a decent living and have a healthy and wealthier life. Therefore, the quantity and quality of a household's human resources shape its livelihood outcomes. Illness and death experienced in a household affect a household's main asset, namely labour, and therefore affect the other assets and resources, and may jeopardize the household's survival and welfare (Karuhanga, 2008). According to Murray (2000), in this way, a livelihood approach enables one to see that the focus is more on agency, actors' capability and their strength, rather than on their needs. In short, agency depends on people's capacity to make choices and take control over their actions during their livelihood generation. According to Karuhanga (2008: 59), 'besides knowledge, skills and good health, people's capabilities also entail the wealth of experience acquired over the years, the ability to negotiate, create and maintain social networks (associability) as well as the potential to improve these attributes'.

3.8 Food security

Since a couple of decades, food security is a global concern. The literature on food security presents various definitions of food and nutrition security. From 1950 on, the challenge was to realize food self-reliance on a global scale. In 1970, through the World Food Conference in Rome, food security came to be considered as a national and world food supplies issue. In the 1970s and 1980s, issues of availability and stable access to food became the key words of the food security problem. Following up on the seminal work of Sen (1981), in 1992, Maxwell and Frankenberger defined food security as access by all people at all times to the food needed for an active and healthy life. Therefore, food security at the household level becomes 'access by households at all times to enough food for an active and healthy life of the household members' (Niehof, 1999: 107). In short, access to food and healthy life became key words, making food security not only a matter of quantity, but also a qualitative issue. Nutrition security requires an adequate dietary intake. Food and nutrition security together refer to food availability, and to the components of the food in terms of calories and micronutrients that are necessary for good health.

The most widely used and accepted definition of food security is the one adopted at the World Food Summit in 1996: 'Food security, at the individual, household, national regional and global levels is achieved when all people, at all times, have physical and economic access to sufficient, safe and nutritious food to meet their dietary needs and food preferences for an active and healthy life' (FAO, 1996). The interplay between food production, food availability, food distribution, and access shape the way in which individuals and households meet their food needs. A household's decision-making power and control over household assets and resources may impact on the access to food within the household, i.e. among household members (Maxwell and Frankenberger, 1992; Niehof, 2010). Still, most often, women are the victims of unequal distribution in households headed by men. In many African cultures, women are supposed to eat only after the men have eaten. However, since the household's food security is a woman's major concern, due to their gender role, women are always concerned about its safeguarding and will initiate coping strategies when their household's food security is threatened (Niehof, 1999, 2003b, 2010).

People cannot achieve food security without access to resources and assets. Poverty increases households' vulnerability, because it affects their livelihood and food security in a vicious circle. Natural resources' depletion, due to the population growth, increases food insecurity (Dietz, 2000). In an economic development context, a households' food security determines its capacity to produce and invest (IFAD, 2001).

Food security on the basis of agricultural production, like in this study area, implies women's access to two very important assets. The first is land, and the second one is labour. We shall see that access to land is regulated by kinship. The second crucial asset is labour. This study shows how agricultural labour shortages may increase a household's livelihood and food insecurity, especially when women have sick household members or relatives to look after. Therefore, to get a reliable picture, it was necessary to obtain information on the main crops produced, farming skills and practices, access to land for crop production, access to agricultural inputs and extension services, and food habits among the autochthonous Agni and the migrants. To assess food availability, I looked at food shortages experienced by households during the past two years, the quantity of food stored by households, the number of meals per day for household members, and the composition of the household's daily menu in terms of protein and food commodities. Focus Group Discussions with men and women provided insights on their perceptions and interpretations of food security, as well as on their coping strategies to deal with food shortages.

3.9 Forest degradation

For some decades now, forest degradation has not only been a global concern, but it is also very much an issue in Côte d'Ivoire (Fairhead and Leach, 1998). The global ecological crisis has placed questions of natural capital and resources degradation and depletion, among which forest degradation, at the forefront of the international development agendas.

Forest degradation is seen as a cause and result of climate change. According to ecologists, forests regulate the climate and water resources. They are the foundation of human, animal, and plant life, and provide food, wood, medicines, and a spiritual frame, for the well-being of humanity. Forests absorb carbon dioxide for growing. Therefore, when forests are burnt through forest fires, or the cutting of trees for firewood, the carbon dioxide and methane (greenhouse gasses) are released and pollute the air. This process motivates the global warming that is continuing to jeopardize a sustainable environment and development (http://www.greenpeace.org/international/campaigns/climate-change/science/forestdegradation, accessed on 08-12-09). The loss of forests implies a loss of their capacity to function as regulators of the environment. This increases flood and erosion hazards, reduces soil fertility, and impacts on the sustainability of human, animal and plant life (FAO, 2000).

People world-wide are faced with environmental problems and the related socio-economic impacts, illustrated through forest degradation, land degradation, air pollution, water shortage and contamination, and the loss of biodiversity (Van den Top, 1998). Eighty percent of the people, mostly low-income groups, rely on natural resources for their livelihoods or for their basic needs. For this reason, they more frequently experience the impacts of poverty related to natural resource degradation. Many studies point out a consensus about the interplay between poverty, population growth and environmental degradation. According to Duraiappah (1996), there is a consensus on the main factors that have contributed to forest degradation in developing countries: logging activities, agricultural and pastoral encroachment and expansion, as well as fuel wood collection. Some studies found that commercial logging activities are the main activities responsible for unsustainable forest degradation in many developing countries (Anderson, 1989; Repetto, 1990; Somanathan, 1991). Other studies point out, on the other hand, that agricultural and pastoral encroachment, motivated by logging trails, cash crops cultivation, shifting cultivators and unplanned settlement, is the key contributor to tropical forest degradation in developing countries (Cruz and Gills, 1990; Westoby, 1991; Davidson (1993). These studies point out the global market forces behind logging activities and cash crop cultivation, driven by the incentives to supply national state revenues in response to foreign demands. Some scholars believe that population growth is the leading factor of forest degradation, as it motivates an increasing agricultural expansion and unsustainable agricultural activities, that aggravate soil deterioration (Southgate 1988; Homer-Dixon, 1999).

Hartmann (2001) disputes the causal relationship in Homer-Dixon's theory between population growth, natural resources scarcity, and conflicts that depict the lack of development of developing countries. From her perspective, the main reason for the lack of development of these countries and the socio-political instability they are faced with needs to be traced back to colonial policy and science, along with the power relations in a global political economy context. Hartmann (2001) believes that conflicts are not caused by poverty due to resources depletion, but because of the wealth of these resources. For her, 'the fact that poor peasants are responsible for most land degradation and forest degradation in the third world is a truism inherited from the colonial era and more recently popularized by our common future' (Hartmann, 2001: 47). Rather, colonial policy and science gained strength through an environmental policy, programmes and projects still focused on alarmist beliefs, such as those about forest degradation and desertification, as consequences of resources degradation in Africa, even if, scientifically, the proofs of the relation of such phenomena with population growth remain limited. In reality, 'the main cause of forest degradation in developing countries is the demand for wood and paper, and nearly half that wood and three-quarters of that paper are used in industrialized nations. The failure to link the consumption patterns of Northern countries and Southern elites to "local" land use, is a major blind spot in Homer-Dixon's approach' (Hartmann, 2001: 52). In fact, relying heavily on developing countries for natural resources, since the colonial time up to now, multinational corporations and the powerful actors of the international political economy continue to exploit African countries' resources. They do so with the complicity of the states involved, and at the expense of the local people. What is more, when a state refuses to collaborate, it will usually face a rebellion, leading to a generalized conflict within the country that increases people's vulnerability and jeopardizes socio-economic development. According to Hartmann, the cases of Congo and Angola are illustrative. In addition, Hartmann believes that Homer-Dixon also failed to see the relationship between the international arms trade and the military, which increasingly motivates and encourages conflicts in developing countries, and therefore adds to the degradation of local environments. In short, Hartmann thinks that the major reason for resources depletion and degradation in developing countries is not population growth. Instead, it is the global political economy of agribusiness, mining and timber industries and their powerful actors, who are impacting on the sustainability of developing countries and their natural resources.

In the same vein, Fairhead and Leach (1998) have criticized the lack of consideration of mainstream environmental policy *vis-à-vis* the local ecological knowledge and practices, which in reality contribute to managing and sustaining the environment. Poor peasants are generally accused of being degraders of their environment, and incapable of being agents of its restoration. 'People have been blamed for damage which they have not caused and have paid heavily for this in policies aimed to control their so-called environmental "vandalism", and to remove their control over resources in favour of national and international guardians ... they are treated as strangers in their own, or hosts' ancestral lands, over which "nature" and its national and international guardians have come to claim a right. Forest degradation orthodoxies may, then, be adding unnecessarily to impoverishment, social upheaval and conflict in what is already a poor region. Inhabitants are denied not only their claims and control over valued resources, but also their own understanding of vegetation dynamics and the ecological and social history with which these are entwined' (Fairhead and Leach, 1998: 192).

The negative discourse of the colonial vision *vis-à-vis* autochthonous knowledge is strengthened through environmental policy, which imposes on local people the adoption of modern practices for environmental regeneration. Instead, the continual failures of environmental policy programmes and projects in developing countries render such policy irrelevant. They stress the need for a critical look at the statistics on forest loss in developing countries, particularly with regard to West Africa, where forest degradation is said to be significantly pronounced. These statistics are incorrect panel data, as they failed to understand and consider the local farming practices and forests conservation practices, which to some extent contribute and sustain forest regeneration mechanisms. By implication, environmental policies and conservation strategies are drawn from erroneous data. Taking the case of Côte d'Ivoire in West Africa, they argue that 'Following FAO/UNEP's forest resource assessment for 1980 (FAO, 1981), Côte d'Ivoire acquired the reputation of having the highest rate of forest degradation in the tropics. The case of Côte d'Ivoire also begins to show how dominant views of forest degradation have obscured counter-evidence concerning past populations and land use and the nature of past vegetation cover' (Fairhead and Leach, 1998: 22). In fact, we cannot deny local people their knowledge, related to the history of their environment, its degradation's interpretation, and the local practices of environmental sustainability. With the economic and cultural importance of environment in their lives, environmental sustainability has become imperative and vital for them. They rely on the environment for their livelihood and cultural rituals. Local ecological practice helps them to overcome forest fires, soils erosion and so on. Discussing the anthropogenic influences on vegetation, Fairhead and Leach (1998: 186-187) believe that 'the enduring effect of settlement on vegetation is fundamental. Inhabitants in their everyday lives can render soils highly fertile, and manage village peripheries in ways which shape tree cover'.

In this study, people in Y.F. saw forest degradation as part of the vulnerability context, relating it to the lack of good soils that impacts on food productivity. It has been apparent that forest degradation has economic impacts, since people have less food production to rely on for the household's consumption, and get less income from selling their produces in the market. In this study, I will look at the vulnerability caused by forest degradation and at local people's conception and interpretation of forest degradation. I will also examine the power relations between local people, the forestry and agricultural extension officers, and the local ecological knowledge on forest protection.

3.10 Research framework

Based on the above theoretical discussions, the research framework presents the vulnerability context of livelihood and food security. The arrows within the framework (Figure 3.1) show dynamic relationships.

Taking a gender perspective, the model in Figure 3.1 shows how the vulnerability context of livelihood and food security is experienced by women, both internally and externally. Women experience internal vulnerability when they lack access to assets and resources at the household level, but also when their matrilineal and patrilineal kinship norms and rules pose a potential risk of increasing their exposure to AIDS. They experience internal vulnerability related to AIDS morbidity and mortality because of agricultural labour shortages, due to the burden of caring for the sick, and, more importantly, in the Agni community due to the large burden of attending too many funerals. The decrease in food productivity that women are faced with is due to labour shortages, exacerbated by the lack of good soils and forest depletion.

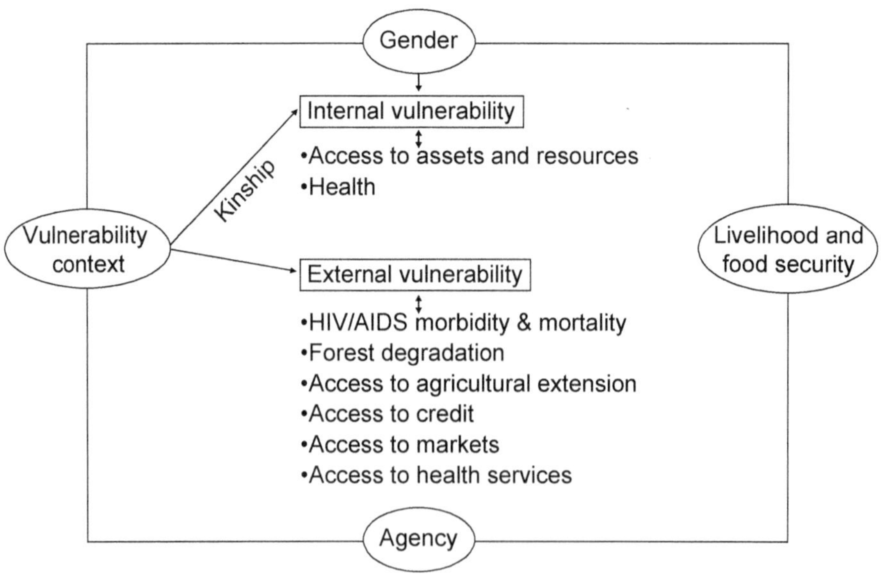

Figure 3.1. Research framework.

Women experience external vulnerability with the lack of access to and quality of institutional structures and social services, such as extension services, credit, markets, and health facilities, which in turn aggravate the internal vulnerability and jeopardize sustainable food and livelihood security. Furthermore, the framework shows how women challenge the economic hardship, while they use their agency to enhance their resilience and improve their position within the social arena. However, this may be done in such a way that it actually increases their exposure to the disease, which may increase their household's food and livelihood insecurity.

Chapter 4.
Research design and methodology

This chapter presents the research design and data collection methods I have used to answer the research questions. It starts with a description of the research design and some methodological issues, i.e. the different phases and methods of data collection and data analysis. The chapter concludes with a discussion on ethical considerations and the problems encountered during the study.

4.1 Phases of the fieldwork

A reconnaissance visit conducted in May-June 2006 started the first phase of the fieldwork. The objectives were to explore the research area, to get secondary data on food production, the AIDS situation, and cultural and gender issues; as well as to get people's views about the problems discussed in this thesis, and to establish contacts with key informants. This field visit started with a meeting with the AIDS programme team of the national institute of rural development (ANADER). The aim was to introduce myself to the team, and the topic I was working on, in order to solicit their involvement and collaboration in the fieldwork phases. At this meeting, we came up with the idea of changing the study area presented at the proposal writing stage, since this area was not part of the top AIDS-affected areas in Côte d'Ivoire. Therefore, we agreed to identify a new study area, based on criteria such as AIDS prevalence and an ongoing implementation of an AIDS project. In line with the above, I visited three study areas: D.B. in the Alepe district in south Côte d'Ivoire, and Z.Z., and Y.F. in the district of Abengourou in eastern Côte d'Ivoire.

During the reconnaissance visit, the main constraints identified in the areas were health problems, AIDS, forest degradation, food and livelihood insecurity, unemployment, and problems with the marketing of food products. Interviews with ANADER extension agents provided information on the agricultural profile of the areas, crop production and the seasonal calendar.

A visit was made to the Alepe district hospital, and to the Abengourou district centre of tuberculosis and HIV testing. Even though both hospitals do testing, I found the head of the Alepe district hospital less inclined to cooperate in the way I had hoped. Presumably, he did not wish to violate the confidentiality of the data on HIV-positive individuals. Although this position is understandable from a professional standpoint, it posed a dilemma for me because in order to carry out my research, I needed to rely on the health centre data about households or individuals who were known to be affected.

Fortunately, I was more successful in the Abengourou district. Unlike the situation in the Alepe district, the Abengourou district has a Voluntary Counselling and Testing Centre (VCT), which is quite big and operational. The VCT, located in the district's tuberculosis centre, is a care unit that coordinates AIDS testing, counselling, sensitization, and awareness activities in the district. The center provides free anti-retroviral drugs (ARVs) with the support of the Global Funds, EGPAF, and PEPFAR (see Chapter 2). Some projects and sensitization campaigns are done in the villages around Abengourou, for testing and counseling, and to generate AIDS awareness. The target groups include people infected with sexually transmitted infections (STIs), people infected with tuberculosis, commercial sex workers, and expectant mothers. During my visits to the center, the staff appeared interested, and interesting issues came up during the discussions I had with them. They stated, for example, that the spreading of AIDS in the district is partly aggravated by cultural attitudes and social drivers, such as stigmatization and exclusion. Those who are tested positive do not respect the medical protocols, out of fear that their fellow villagers see them visiting the hospital to have their treatment. They fear the stigmatization and exclusion that are likely to be the result. The principal doctor and head of the tuberculosis and HIV testing center of Abengourou appeared to be very interested in my research. He perfectly understood my dilemma of identifying affected individuals and households without explicitly targeting them. He found a most creative solution to keep his medical secret on the one hand and to direct me to affected households on the other. He proposed that I would be assisted by his principal helper, who knew the village and its people very well. He would see to it that affected households would become included in my survey in an anonymous way, so as not to compromise confidentiality. Thus I was allowed to make use, in an indirect way, of the data from the center. It was agreed that the head of the center would guarantee an ethically correct use of the data by becoming my advisor and local supervisor for the fieldwork. During the first phase of the fieldwork, we agreed to select Y.F. as the study area, for several reasons. Not only is this village located in the second most AIDS-affected district of the country, but the village is also a pioneer area for AIDS awareness projects in the region, and well known to the staff of the VCT. Finally, it is also an important area in terms of food production. Its history as a prominent cocoa production area has attracted many migrants, which provided an opportunity for studying the effect of cultural factors, like matrilineal and patrilineal inheritance rules, on livelihood and food security.

The second phase of the fieldwork started in January 2007 and lasted until April 2008. It involved the introduction of the research to the district officials, and the general and qualitative data collection on the situation with regard to AIDS, food security, and forest degradation in the area. The selection of field assistants was done during this phase as well. I also used it to operationalize the key concepts and identify other variables that should become included in the survey questionnaire, and to develop a household survey plan with the village leaders. A household survey was conducted in the third phase of the fieldwork. It was carried out between April and July 2007. A preliminary analysis was done to identify and fill the gaps. Finally, I used the last phase of the fieldwork, from August 2007 to May 2008, to obtain qualitative data in conjunction with the quantitative survey data through in-depth interviews, cases studies, FGDs, and key informant interviews. During this phase, discussions on AIDS in Côte d'Ivoire and the government's response to the epidemic were held in the capital of Abidjan with officials from the Ministry of the Response to AIDS (MORA) and the Ministry of Public Health (MOPH); with officials from the Ministry of Agriculture (MOA) I discussed the food security situation related to AIDS in the country; with officials from SODEFOR, the National Forestry Agency and with the National Institute of Rural Development (ANADER) officials and extension agents, I discussed the trends in crop production over the last few years.

4.2 The study design

The study is designed by a mixed methods approach that combines both quantitative and qualitative methods for explanatory and descriptive purposes. The mixed methods approach allows using different methods to collect data, with the possibility to assess, to compare, and to validate the information. Many scientists have shown the importance of using the mixed methods approach (Green *et al.*, 1989; Scrimshaw, 1990; Ritchie, 2003). For instance, Green *et al.* (1989) mention triangulation, complementarity, development, initiation, and expansion as the five ways of applying a mixed methods research strategy. For triangulation, I used different methods (like key informant interviews, survey data, and Focus Group Discussions), in order to collect data with the intention of comparing different aspects of the same topic. For example, on the issue of household food security, I compared data on crop production with those on the daily food intake and the distribution of food to members of the household. Using data in this complementary manner helped me to verify and validate the correctness of the information.

I used qualitative research methods to fine-tune the concepts and questions for the quantitative survey. In-depth interviews helped to go deep into some outcomes of the survey, so as to grasp the understanding and standpoint of the participants. Actually, these interviews helped not only to contextualize, but also to interpret participants' perspectives about issues such as gender relations in agricultural labour, resource allocation, matrilineal and patrilineal kinship rules and practices, and the impacts of AIDS and other problems, such as forest degradation, on households' livelihoods and food security. The different methods of data collection and their role in the study are summarized in Table 4.1.

Table 4.1. Mix of methods used in data collection according to phase of fieldwork.

Fieldwork phase	Methods	Emphasis
1st phase May-June 2006	Reconnaissance visit: interviews with key informants, consulting secondary sources, discussions with local NGO	Getting key local actors committed to the research plan, exploration of the situation in the area regarding the research topic, selection of research sites
2nd phase January 2007-April 2008	Selection and training of research assistants, interviews with key informants, participant observation	Introducing the research to key local actors and officials, operationalisation of the key concepts, identifying relevant variables for the questionnaire, preparing the survey
3rd phase April-July 2007	Household survey	Quantitative data collection on the key issues
4th phase August 2007-May 2008	Key informant interviews, in-depth interviews, case studies and Focus Group Discussions	Validating the survey data, in-depth follow-up of the survey, filling gaps

4.3 The methods of data collection and analysis

The study design combines qualitative and quantitative research methods. Data collection techniques included a secondary data review, a household survey, Focus Group Discussions, in-depth and key informant interviews, case studies, and participant observation.

4.3.1 Secondary data

Relevant reports and documents from the Ministries and organizations in Abidjan and the Abengourou district, as well as from literature and internet sources, provided secondary data on the physical and demographic characteristics of the study area, kinship systems and cultural landscapes, crop production and farming systems, AIDS, food security, the problem of forest degradation, gender and development issues, as well as existing AIDS mitigation policies and programmes.

4.3.2 The household survey

In this study, the household is used as the basic unit of analysis (see Section 3.2). According to the conventional definition of a household, a person or group of persons are considered as members of a household, based on criteria such as living under a 'roof' or within the same compound or homestead for a continuous period of at least six months of the past year, sharing food from a common source, and contributing to or sharing in a common resource pool. However, this did not match with the criteria for determining household membership in the research area. In the field, the household may include a male household head who regularly migrates to support the household financially in its daily needs. Therefore, even when migrated, men still are the major decision-makers, involved in decisions such as the children's education, marriage and burial ceremonies, as well as the sale or purchase of the household's major assets. For this reason, in the field, I used the *court* (compound) as an operational definition of household. In the study area, the household size is about eight, but some very large households are found among the migrants, counting up to 35 persons.

A survey questionnaire was administered to 191 households. The questionnaires were administered face-to-face, in a friendly atmosphere. The male head and female heads of the household were interviewed, including *de facto* female heads of household, because of sick or migrated men. Data were collected on the household's structure, assets and resources, livelihoods strategies and labour allocation, crop production, the effects of AIDS and forest degradation on household assets, livelihood activities, and food security.

The sampling procedure

At the writing stage of the proposal, approximately 200 households were considered to form a representative sample. It included the autochthonous Agni and migrants' households and AIDS-affected as well as non-affected households. In the end, the effective sample size was 191 (Table 4.2).

I used the following criteria as proxy indicators for being affected: affected households were considered to be households in which a member who had been contributing cash or labour to the household was experiencing illness due to chronic malaria, called *babasso* in the local language, tuberculosis, pneumonia, meningitis and/or typhoid, diabetes; when the illness had been lasting for more than six months and had been recurrent; or when a household member had died during a period of five

Table 4.2. Sample household survey (n=191) according to AIDS status.

Households	AIDS status		
	Affected	Non-affected	Total
Agni	63	39	102
Migrants	52	37	89
Total	115	76	191

years prior to the time of the survey of any the abovementioned diseases. In addition, indications were a household member who had either moved in or out after the death of a relative in an affected household; and a household fostering orphans whose parents were ill or had died from the abovementioned diseases. The last criterion was households with children withdrawn from school, due to morbidity or mortality in the household. Non-affected households were identified to be households that had not been affected by illness or death of a family or household member, and that were not fostering AIDS orphans.

The training of interviewers

The selection of interviewers has been facilitated by the local leaders of the village. Two supervisors and 10 interviewers were selected based on their education level, working experience in development projects, fluency in spoken and written French, and a good knowledge of both the Agni language and Bambara, which is the language spoken by most of the migrants' communities in the area. All interviewers were inhabitants of the village. The supervisors and the interviewers got a training comprising an explanation of the goals and objectives of the study, as well as specific skills to motivate the respondents, and the translation of the questionnaire into local languages. Additionally, the interviewers were asked to keep the information confidential. Each day after the interviews, the researcher and the supervisors went through the questionnaires with the interviewers to find out whether there were gaps or unclear responses from the respondents. Interviewers with many unanswered questions went back to the field, to complete the questionnaire.

Although, at first, the interviewers and respondents found the questionnaire too long, after the study's importance for the village and the country was explained, they were motivated to support the study. The respondents expressed the wish that the findings of the study be used for development projects in Y.F.

Topics addressed in the survey

During the household survey, data were collected on: households' characteristics, the assets and resources used; households' livelihoods activities, households' food production and consumption, the effects of illness or death, and of environmental degradation on livelihoods activities and food security (see Table A1 of the Appendix).

Data processing and analysis

The survey data were processed with manual editing and coding, and analyzed using the SPSS statistical package. This provided descriptive statistics, which enable the connections between different variables. From the index variables constructed, independent variables (AIDS-affected/non-affected, food adequacy) and dependent variables (livelihoods assets, activities, and incomes) were enabled for comparison. The formulation of proxy indicators led to the distinguishing of AIDS-affected households from non-affected households. The household asset value was assessed based on household asset ownership and the value of each tangible asset.

In addition to the quantitative methods, a number of qualitative methods were used.

4.3.3 Focus Group Discussions (FGD)

Focus Group Discussions were used to understand the nature of the problem discussed in this thesis, by eliciting information from discussions on a particular topic between a small group of participants who share a common interest or position, and to learn about their experiences and perspectives. Data were generated and enriched by the interaction of the participants among themselves, as observed by the researcher. This led to the identification of people's needs, perceptions, and expectations. Participants of the FGDs were selected on the basis of their agricultural activities, village leadership, membership of women's organizations and community-based organizations that respond to AIDS or engage in forest protection. Based on gender issues and cultural differences, we decided to form separate focus groups of Agni men and Agni women, as well as of migrant men and migrant women, aiming at gaining insights in gender relations in the different cultures, and the impacts of this on households' livelihoods and food security in the context of the AIDS epidemic. Additionally, we formed a mixed group composed of Agni and migrant youths, to get their perspective on the issues discussed in this thesis, their relations with the elder generation, and their expectations in terms of responding efficiently to the devastating impacts of AIDS in order to achieve a sustainable development of the village. In short, five FGDs were conducted, their size varying from eight to sixteen people (Table 4.3). The segregation of the groups by gender and by kinship system was aimed to provide an environment for free expression. The selection of the FGDs has been facilitated by the local leaders. All the FGDs were moderated by the researcher with the help of the field assistants, who in addition acted as translators for some local words and expressions. The FGDs lasted for 60 to 100 minutes. The discussions were held in the village at different places, at the convenience of the participants, and were audio-recorded. The topics discussed during the FGDs included gender-related livelihood activities; resource allocation; the availability of time; agricultural labour allocation; matrilineal and patrilineal land inheritance; cultural norms and taboos; sexuality; health/AIDS problems and programmes; forest degradation; food security; community integration; coping strategies; and the challenges of overcoming vulnerability. A list of topics can be found in the Appendix.

Table 4.3. Participants in FGDs (n=59).

Focus group	Male	Female	Number of participants
Group of Agni men	8	0	8
Group of migrant men	9	0	9
Group of Agni women	0	14	14
Group of migrant women	0	16	16
Group of Agni and migrant youths	12	0	12
Total	29	30	59

4.3.4 Key informant interviews

Key informant interviews were a strong help to get individuals' perspectives about the nature of the problems discussed in this thesis. Key informants were selected with regard to their experience and professional background, as well as their position and activities. They are knowledgeable about people's livelihood activities, cultural systems and norms regarding communities, food policy and the food security situation, the environmental degradation problem, the AIDS epidemic, and impacts in the area. The 18 individuals selected as key informants included officials working for the Ministries of Agriculture, Public Health, Response to AIDS, the Forestry Agency, and agricultural extension providers; among them also were district administrative officers, local leaders, and workers in communities-based organizations. The interviews were conducted using guides specific to the category to which the informant belonged. However, they were all asked questions about perceptions of food security issues in relation to gender and AIDS. A list of key informants interviewed can be found in the Appendix.

4.3.5 Case studies and observation

Using an actor-oriented and ethnographic approach, I investigated particular real-life situations. This required me to observe and describe social situations, and the actors' interactions and social constructions of responses to vulnerability related to livelihood and food insecurity. Ethnographic experiences underpinned the choice of social theories I used to understand empirical reality. 'Case studies serve to establish the validity of a particular theoretical principle, not by achieving statistical significance but through their ability to elaborate a theoretical principle, by confronting it with the complexity of empirical reality' (Mitchell, 1983 cited by Karuhanga, 2008: 102). As stated by Burawoy (1998: 16), with ethnography 'we begin with our favorite theory, but seek not confirmations, but refutations that inspire us to deepen that theory'. Ethnographic research leads to challenge dominant modes of thinking (such as the AIDS mitigation policy in Africa), and 'brings to light usually disavowed, ignored, or obscured empirical realities. It contributes to making thinkable what was unthinkable. Whereas in the current political and economic order, certain views on life are constantly being disabled and excluded, ethnography provides the chance to disclose different realities and political possibilities' (Koster, 2009: 30).

I used ethnographical cases to contextualize and capture a generalized understanding of social processes, and to explain particular social situations. I carefully examined the events to reconstruct narratives, for example about the effects of AIDS and forest degradation on food security in the eyes of the women themselves. Subsequently, I used observation to triangulate the information collected with other methods, particularly regarding women's labour reallocation, and the households' daily food intake and food security. For example, while the survey data stressed that, in the sampled households, people eat three times a day, this was clearly contradicted and qualified both during FGDs and observation, with regard to what was seen as a proper meal or, rather, as the quality of food. Apparently, women's pride does not allow them to disclose their precarious status regarding their household's food security. This was also reported by Nombo (2007) and Niehof (2010).

Through case studies of affected households, I examined the ways in which AIDS and other adversity, like forest degradation, influence livelihood activities, gender inequalities, increase vulnerability, and jeopardize any sustainable development. Through case studies, observation and conversation, and through discussions during informal visits to the communities, I gained a better understanding of issues such as culture, gender relations, people's attitudes towards AIDS and forest degradation, and the effects on food security, as well as the coping strategies women use in meeting the challenges posed by the external vulnerability context. In this study, there are several case studies of which the focus slightly differs. In some of them, the focus was on access to resources such as land, labour allocation, labour shortages related to AIDS morbidity and mortality, households' livelihood strategies and food security; while in others, the focus was on people's interpretation, their knowledge and narratives on the AIDS epidemic and its impacts on livelihood and food security, forest degradation, and women's battle to challenge their vulnerability. The selection of cases has been facilitated by local leaders, based on criteria such as AIDS-affectedness, membership of women organizations, of a *komian* organization, involvement in activities to respond to AIDS, and forest protection activities. A checklist guided in-depth interviews for the case studies as well as for critical observations. The notes were recorded in a field notebook.

4.4 Ethical considerations and problems encountered during data collection

Like in any AIDS impacts research, I encountered some methodological and ethical challenges during this study. The stigma and silence related to AIDS rendered it difficult to identify households that have members who are currently ill or have died from AIDS, because people are usually not willing to admit or to know their HIV status. Therefore, as in most AIDS impact studies, I had to rely on proxy indicators to identify AIDS-affected households. However, as stated by Wiegers (2008: 181), proxy indicators have several limitations, because 'proxy indicators and verbal autopsy means researchers are attaching labels to diseases or causes of death outside people's knowledge'. It leads to a bias in the researcher's interpretation, since not all chronic illness or premature death is due to HIV/AIDS. Although we got guidance from the health center when trying to gain insights on affected households, we still have to doubt the information given by some respondents, since they did not want to provide data relating to their HIV-status. Even more, this raised ethical concerns, because households did not give their consent to the health centre to share their status with others and some of them might not even know their own status.

Furthermore, I experienced difficulties to measure the impacts of AIDS on households' food and livelihoods security. Narratives are based on a number of assumptions that are difficult to measure. For example, while in the household survey, people say they eat three times a day, during the FGDs they acknowledged that for breakfast and lunch they have snacks, rather than a proper meal, because they cannot afford the latter. People also felt ashamed to talk about their household food insecurity during the household survey. Wiegers (2008) pointed out these complex and challenging issues in AIDS impacts research, related to the measurement of the impacts of AIDS on rural livelihoods.

I concur with Wiegers (2008: 182) that 'AIDS interacts with a complex set of environmental, socio-economical and political factors which tend to differ per location and the effects of which are hard to disentangle from the effects caused by the epidemic'.

Chapter 5.
Internal vulnerability of Agni and migrant households

This chapter examines the ways in which women experience and respond to vulnerability at the household level. As said, vulnerability has an internal and external dimension. At a micro level, the vulnerability lies in the constraints people experience in their daily lives and the strategies they develop to generate their livelihoods. Results from the survey research, the five FGDs, interviews with Agni and migrant local leaders, key informants, the case studies, as well as participant observation all demonstrate that women experience internal vulnerability in terms of a lack of access to certain assets and resources in their daily life. Furthermore, results show that cultural norms, practices and stigmatization shaped by matrilineal and patrilineal kinship systems may increase women's vulnerability to AIDS impacts, jeopardizing the household's food and livelihood security. The chapter presents women's livelihoods activities, constraints faced and strategies used in the process of livelihoods generation in an AIDS context. It describes at a micro level, the impacts of AIDS morbidity and mortality on food and livelihoods security, and the ways in which women respond to it.

5.1 Profiles of Agni and migrant households

This section presents both Agni and migrant households' profiles in terms of demographic figures. From survey research, using indicators such as AIDS affection, household size, education, marital status, ownership of tangible assets, access to and use of water and energy, as well as land ownership and use right, its becomes apparent that both Agni and migrant households experience internal vulnerability for different cultural reasons and with different social effects.

5.1.1 Demographic characteristics of households

Out of the 191 households, 102 are Agni households and 89 are migrant households, while 60% (n = 115) are HIV-affected and 40% (n = 76) are unaffected households. About 55% of the affected households (n = 63) are Agni households, and 45% (n = 52) are migrant households.

The distribution of household headship according to gender shows 74% to be male-headed (n = 141) and 26% to be female-headed households (n = 50). In terms of ethnicity, the data show that there are more women heads of household among the Agni than among the migrant households: 86% (n = 43) and 14% (n = 7), respectively. The average age of the household head is 42 years. The respondent were also of various religious backgrounds, with Catholics (60%), Protestants (10%) and Muslims (30%) being the dominant religious categories.

In terms of the level of education, 54% of the respondents have attained a primary education, 19% have a secondary and college education, and 27% have had no formal education. Women are less educated than men: among those who have attained primary school, 61% are men while 39% are women. Among those who have attained a secondary and college education, 72% are men, while 28% are women. In terms of ethnicity, Agni men and women are generally more educated than migrant men and women. Seventy-two percent of the migrant men had no formal education, versus 28% of the Agni men. Seventy-five percent of the migrant women had no formal education, against 25% for the Agni women. These results reflect the general education patterns in Côte d'Ivoire[14], with gender disparities pronounced among the migrants, who mostly come from the northern part of the country and from neighbouring countries (see Chapter 2). Most of them are illiterate and live in the host villages where they work as agricultural workers.

According to the respondent's marital status by household headship, out of 43 households headed by Agni women, 18 have divorced and are separated, and 5 are widows. In contrast, migrant women heads had no divorced, separated or widowed status. However, polygamy is more pronounced in migrant households than it is in Agni households, with 22 cases out of 82 households for migrant men and three cases out of 59 households for Agni men. From interviews and participant observation, it came up that migrant women rarely are single, divorced or widowed compared to Agni women, because of the fear of stigmatization when one remains single. Rather, they are willing and manage to get remarried after a failed marriage or after widowhood, as culturally recommended. From a cultural perspective, an unmarried woman is not well respected. On the other hand, Agni women do not mind being divorced or a widow. During FGDs with Agni women, it transpired that Agni men are unwilling to marry divorced or widowed Agni women. They prefer to get married to young women without a background of divorce or widowhood. However, according to Agni men, this has to do with the pride of Agni women, who mostly want to show their capabilities in taking over the household headship after a failed marriage or widowhood.

The issue of the marital status of Agni women has a bearing on their household's socio-economic vulnerability. As stated by a participant during the FGDs with Agni women: *'when an Agni man divorces you, he divorces you with your children'*. In other words, divorced Agni women face the burden and workload of the household headship alone, without any contribution from the ex-husband, i.e. the father of the

[14] These figures are well below the 2008 national census statistics that give 70% of the males and 52% of the females of the population aged 15 to 24 to have attended school.

children. From participant observation, I witnessed the difficulties Agni women are going through as heads of their households. Being *de facto* single, without a formal partner, increases their burden and the economic hardship they are faced with. This potentially exposes some of them to transactional sex and HIV, in which they engage to be able to take care of their children.

In terms of the distribution of household size, Table 5.1 shows that most of the households have 5 to 9 members; that is the case for 52 Agni households and 42 migrant households. Some larger households are found among the migrants, because they have more polygamous households. The average household size is about 8 (Mean = 8.08).

5.1.2 Tangible assets of households

Survey data show that in the area, households' main tangible assets are mainly a bicycle and a radio. Few households own a television set. Many households own a mobile phone. These assets belong mostly to men. In terms of farm equipment, households own hoes (*daba*), machetes and sickles. Most of the sampled households (90%) have permanent houses. I was surprised to observe some villas with air conditioning in the village. However, this was related to the fact that the setting has known a prosperous period in the past (see Chapter 2).

Survey results show that 57% of the households possess poultry and small livestock, mainly goats and sheep. Most of these households complain about the depletion of these assets, which they have sold to cope with AIDS-related care as well as the burden of funeral expenses or financial contributions to assist the bereaved families

Table 5.1. Household size according to ethnicity (n=191).

Number of household members	Agni		Migrant		Total	
	n	%	n	%	n	%
1-4	18	17.7	21	23.6	39	20.4
5-9	52	51.0	42	47.2	94	49.2
10-14	24	23.5	17	19.1	41	21.5
15-19	8	7.8	4	4.5	12	6.3
>20	0	0.0	5	5.6	5	2.6
Total	102	100.0	89	100.0	191	100.0
Mean HH size	7.88		8.31		8.08	

Household Survey, 2007.

in the village. This is in line with Karuhanga's findings (2008: 264) that, in an AIDS context, 'increased expenditures needs associated with AIDS-related care and funeral expenses necessitate household resources to be converted to cash'.

5.1.3 Water and energy access and use

The houses are connected to an electricity network. From participant observation, I realized that because of the poverty in the village, people make arrangements to share the electricity connection and the bill. The major source of energy for cooking and heating water remains firewood, which women collect from the fields. The women complained about the scarcity of firewood in the forest. From the FGDs with Agni and migrant women, it came up that the scarcity of firewood in the forest is due to forest degradation. Based on observation, I gathered that households are now using other species of wood than they did in the past.

The sources of water used by households for drinking, bathing and washing clothes are well water and piped water. Almost 35% of the households have to fetch water far from their houses. The FGDs with Agni and migrant women showed that women have to fetch and carry water to the house every day. According to them, this is not only tiring, but also time-consuming. Survey data indicate that, on average, they make seven trips per day to fetch water from the public fountain, located in the village center. Firewood is collected from fields, approximately five times a week, which takes two hours per trip. Women are helped by girls in performing these domestic activities. The type of latrine most commonly used by the households is a pit latrine, i.e. a traditional one, located mostly behind the house.

5.1.4 Land access and use

Survey data about land size, quality, use and tenure, stress that land belongs exclusively to the autochthonous Agni. On average, the total agricultural land owned by a household is about three acres. Unlike some other rural settings in Côte d'Ivoire, in this village, migrants do not own land. Only two migrant men reported to have gotten land from the chief of the village because of a strong friendship their late fathers had with the chief, but this is an exception. Usually, in this setting, migrants acquire land for crop production by renting or sharecropping with the land owners, i.e. the indigenous Agni, called *tuteur*. This practice is widely known throughout the cocoa-producing region of the Akan people, as it is among the Baoulé (Van den Breemer, 1984).

From a gender perspective, it has emerged that women and men do not have equal land rights. Results show that Agni women get land rights in their households through inheritance and through marriage, but also through the permission to use plots from their Agni relatives (Figure 5.1).

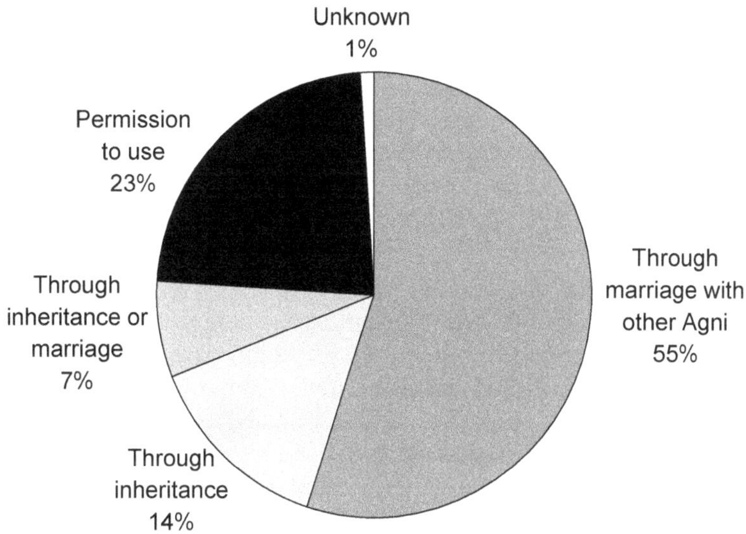

Figure 5.1. Agni women's access to land. Source: Household Survey (2007).

Regarding land use rights in Agni culture, there are multiple narratives on women's access to land. From the FGDs with Agni men, it became apparent that women do not have the same land use rights as men. From the men's perspective, land definitely belongs to men, and can be inherited only from maternal uncles. However, results from the FGD with Agni women show that they can inherit land in their matrilineal group when there is no legal inheritor, i.e. when there is no living male in the matrilineal group. Accordingly, when they inherit land in such circumstances, they pass it on to their sons. So, the high percentage of women heads of household in the Agni cluster (43 households out of 50) is possible because of opportunities they may have to get access to land, even after becoming divorced or widowed. An example, Case 5.1, shows an Agni woman who thus has inherited land from her relatives.

Case 5.1 Land ownership and access for an Agni woman

Cathi is a 52 year-old divorced woman born in Y.F. village, in a family of three children (one boy and two girls). Her brother and her parents are dead. Cathi has two boys and three girls. She is living in the village, but her two sisters are now living in Abidjan. They regularly visit each other. Her sisters are not interested in farming activities; they are running small businesses in Abidjan. Cathi has to rely only on food crop production to feed her children, without the support of their father. Hence, unlike her sisters, Cathi is concerned about her access to land and her use right. She narrates part of her experience as follows:

> *'I have gotten the land use rights through inheritance. The land belonged to my grandparents, who left it to my mother, because there was no nephew, either from my grandfather's family or from my grandmother's family, to inherit it. Thus, my mother has inherited all the land. When my mother died, I inherited the land together with one brother and two sisters. My brother died, and my sisters are now living in Abidjan, and are not interested in crop production. So, I am now the only one who is using the land' [Cathi, Y.F., 15-02-08].*

From FGDs with Agni women as well as from personal observation, I realized that land ownership is, to some extent, flexible in the Agni community. Furthermore, women can own land when there is no culturally legitimate inheritor, i.e. a nephew like in Cathi's case, but also, unlike in the past, both male and female children can nowadays inherit land from their relatives, even when the nephew is alive. This shift was made possible by a national law promulgated for flexibility in land ownership in Côte d'Ivoire (Law N°64-379 of 7 October 1964), which allows people to choose between the options of patrilineal or matrilineal inheritance. This national law challenges land use rights from a matrilineal system's perspective. In fact, the law emerged due to pressure from powerful autochthonous intellectuals, who opposed the traditional land ownership that denies them the right to inherit land from their patrilineal relatives. However, from secondary data, it has become evident that, regarding land ownership in Côte d'Ivoire, cultural rules are still conflicting with modern law, and that this is affecting women more than men. Indeed, within the Agni community, women are still confronting cultural rules that deny them the right to land ownership.

From the FGDs with Agni men, I realized the weight of gender ideology in their narratives on women's land access and use rights in the Agni community. Even though women in fact do sometimes get land use rights, men do not want to admit it. They refuse to acknowledge the principle of equal access to land use rights. Although women have less formal rights to ownership of land, they have opportunities to have access to land for food production, which is vital for their households' food security.

The land rights situation of migrants differs from the indigenous Agni in two ways. Firstly, the mostly non-Akan migrants acknowledge patrilineal inheritance rights. Secondly, migrants do not have inheritance rights on Agni land and can only get rights to land through marriage or a gift. Survey data, as well as FGDs with both Agni and migrant men and women confirmed that migrants do not have ownership rights to land in an Agni-dominated village. They can only have access to land by renting or sharecropping. This is illustrated by the case below, which narrates of a migrant woman's access to and use of land.

Case 5.2 Land ownership and access for a migrant woman

Sali is a migrant woman coming from Mali, a country bordering on northern Côte d'Ivoire. She is aged 38 and married with 6 children. The children have all been born in the village, and live there now. They are all involved in farming activities with their father, who is engaged in a sharecropping arrangement. Below, Sali relates the migrant woman's experience of being landless.

> *'Unlike for the Djatigui (Agni), our basic food crop is maize. We have to rent land for food production to feed our households, since migrant people are not allowed to own land here. Actually, renting land is very difficult for us, as we have to pay 15,000 or 20,000 francs CFA for the year. Many of the migrant women cannot afford that. Apart from this option, you can get a plot from a land owner and share the production with him. With this option also, we cannot get enough food for our households. So, mostly we help our husbands to produce food with the same options, i.e. renting land or sharecropping' [Sali, Y.F., 15-02-08].*

Being landless, migrant women have to rely more on small business, i.e. they buy food produces such as cassava, banana, yams, vegetables, and fruits from those who farm and make a profit by selling these to people in the surrounding areas. However, difficult access to markets because of the transportation costs makes it hard to get a profit out of this business. For this reason, these livelihood activities do not actually improve their socio-economic situation. Additionally, migrant women are faced with more difficulties feeding their households because their households are larger.

From FGDs with Agni and migrant men and women, and from observation, it became apparent that women hardly participate in household decision-making and have no control over household assets and resources. Agni and migrant men endorse a patriarchal, male domination in the household. They see themselves as *the household chief*. While women do not claim the household chief's position, they fully acknowledge their own key role in the household's food provision and in maintaining its food security.

5.2 Culture, matrilineal and patrilineal kinship, and AIDS

Cultural norms and practices are part of the internal vulnerability that women experience at the household level. From FGDs with Agni and migrant men and women, from interviews with Agni and migrant local leaders, as well as from participant observation, it transpired that cultural norms and practices, shaped by matrilineal and patrilineal kinship systems, may increase women's vulnerability to AIDS.

Below, I will present cultural and kinship practices of the autochthonous Agni and patrilineal migrants, and show how norms and socio-cultural practices, experienced within their households, add to women's vulnerability to AIDS impacts. A case in point is the cultural dilemma of stigmatization.

5.2.1 The culture and matrilineality of autochthonous Agni

From the literature, as well as from my interviews with the chief's spokesman and from the FGDs with Agni men and women, it clearly transpires that the Agni kinship system is matrilineal. Especially the assets and royal/chieftaincy positions are inherited by nephews (the sister's sons) from their maternal uncles (Van den Breemer, 1984; Kofi, 1990). Therefore, kings and chiefs come from the mother's side, i.e. from the children of sisters. In such a context, women do get a certain organizational power (Wolf, 1990), since they are the ones who are supposed to protect Agni culture through their reproductive role, by ensuring the sustainability of traditional political authority. Therefore, in Agni culture women are much motivated to bear children, to ensure the continuity and sustainability of the royal descent group and the community. While performing their reproductive gender role and fulfilling their duty *vis-à-vis* their culture,[15] some women may resort to having several partners in order to produce children, which in the present circumstances may increase their exposure to AIDS. It is common and culturally acknowledged that when married Agni women experience difficulty having children, they are requested to divorce and look for another husband. The other way around, an Agni woman can also request her husband to take another wife when she does not bear children. In the context of the AIDS epidemic, this may increase the risk to contract the disease. Agni women do have control over their sexuality until they get married. Once Agni women are married, their culture does not allow them to refuse sex to their husband. If they do, they are fined to payment of a chicken to their husband.

Although Agni girls are expected to guard their virginity before they get married, this is not as strong a cultural obligation as it is for the migrant girls. One can observe that the Agni house consists of several rooms around a central courtyard. The principal bedrooms for parents are separated from the children's bedrooms, which are usually at another side of the courtyard. Hence, some girls may not be under strict parental control and may have a certain amount of sexual freedom.

[15] For women who belong to the royal families, it is likely that the motivation to bear children is related to more pressure to bear them, in order to provide potential kings and chiefs. In some circumstances, this may lead them to have several partners.

5.2.2 The culture and patrilineality of the migrants

From FGDs with migrant men and women, and from interviews with local migrant leaders, it became clear that, following their patrilineal kinship rules, migrant women are not supposed to have any sexual freedom. Girls and women are not supposed to make decisions about their own sexual life. Before getting married, girls are obliged to keep up their virginity and to honour their families as culturally requested. Although I could observe that this practice is not as strong as it was in the past, nonetheless, married women in a patrilineal migrant culture see themselves as sexually submissive. Women are at the mercy of their husbands, even in case of a suspected HIV infection. Therefore, they accept risking their life, out of respect for their culture and for fear of getting stigmatized.

> *'Mostly, we women are afraid of asking our husbands to use condoms. We cannot. We are noticing here that when a husband dies this year, the wife will follow next year. We are aware of the problem, but we don't have any choice. We are only obliged to support. So, if we are not infected yet, we are the potentially infected. You can suspect that your husband is infected, but what can you do? You know that all of you, the members of your household, will one day be infected and die; you have to leave it up to God' [Kadi, Y.F., 15-02-08].*

> *'Often you are sick, but you are not allowed to refuse sex to your husband, unless you accept to be beaten by him. Most of us are beaten inside the house. Thus, we women should never refuse sex to our husbands, even when we are sick' [Alima, Y.F., 15-02-08].*

From my perspective, women 'do gender', i.e. they play their gender role, since respecting culture has something to do with their womanhood, and the future of their children. In other words, according to the migrants' beliefs, as long as women respect their culture by being at the mercy of their husbands, they and their children are blessed, will have a good future in life, and go to heaven as well. According to this belief, the children's future depends on the submissive behaviour of their mothers *vis-à-vis* their fathers. In other words, *bebi bah boloh*: everyone stands on the hands of his/her mum. Therefore, women can earn respect and improve their position in society only if they are fulfilling their culturally prescribed marital duties, even if they know their husband is HIV-positive. In the latter case, they may risk their life by respecting culture, unless they accept to be stigmatized by the entire community. In contrast, husbands can ignore their wives' sexual desires without any consequences resulting from cultural norms. This is experienced by both migrant women and Agni women.

During the FGDs with migrant women, the point was raised that, in the context of polygamy, many women are living with the fear of stigmatization.

> *'When a woman reaches a certain age, the husband espouses another one and abandons the first, even in bed. You cannot complain, because of stigmatization. Otherwise, people will think that you are not capable of containing your sexual desire. This is shameful for you' [Fati, Y.F., 15-02-08].*

Being incapable of expressing their sexual desires because of stigma may lead some women to look for extra-marital partners to satisfy their sexual desires. As stated during the FGDs with migrant women, 'women who are not strong, sell themselves'. By 'selling themselves', that is, by engaging in sex with extra-marital partners, those women expose themselves, along with their respective households members, to HIV. Additionally, sororate and levirate[16] are cultural practices that are still practiced by migrant people. These practices increase women's exposure to HIV as well.

In short, in both kinship systems, women do experience the weight of gender ideology. On the one hand, there is a strong motivation to bear children in order to sustain the lineage and community; on the other hand, there is the stigma on the expression of sexual desires, for married women which may expose some of them to multiple partners, and therefore to HIV. In addition, from the five FGDs I gathered that polygamy is culturally permitted by both communities. It has the potential of increasing women's vulnerability to HIV and spread the disease among the household members as well. In both cases, married women are not allowed to refuse sex to their husbands, but some women may sometimes negotiate. Otherwise, they are exposed to cultural sanctions. These sanctions are not imposed in the private sphere, but by public exposure of the woman's attitude, making women the subject of strong frustration and shame. Men, on the opposite, can refuse to have sex with their wives without being exposed to sanctions.

In line with the above, both matrilineal and patrilineal systems seem to lead to a feminization of AIDS in both Agni and migrant communities, although in different ways. These systems therefore increase the potential exposure of women to AIDS, and the related impacts on food and livelihood security in rural areas.

[16] Levirate: when a widowed woman has to get married to her brother-in-law. Sororate: when a widowed man has to get married to his sister-in-law. In the AIDS context, it is noteworthy that these cultural practices increase household vulnerability to the disease, by spreading it more rapidly among household members (see Fox, 1967 on kinship and marriage).

5.3 Impacts of morbidity and mortality on livelihoods

The results from the research survey and cases demonstrate that households experience morbidity and mortality related to AIDS and its related impacts on food and livelihoods security. Yet, it has also become apparent that, compared to their male partners, women experience the burden of this internal vulnerability more intensely, as they are the ones faced most with agricultural labour shortages, due to the burden and workload of caring for the sick, and of being at the forefront of funerals.

5.3.1 Impacts of morbidity on livelihoods assets and activities

In the survey, illness and death were mentioned by households as the main causes of the food and livelihoods insecurity they had experienced during the past five years. The respondents were asked whether any member of their households was sick and whether any member of their households had passed away during the past five years. A number of diseases as well as the causes of death were listed by the interviewers. Out of the 115 affected households, 83 households mentioned illness and 32 mentioned death as the main cause of food and livelihoods insecurity. The data reveal that morbidity and mortality affect the households' food and livelihood security considerably (Table 5.2).

Most of the households (54) mentioned duration of an illness period of over six months. Regarding the cases of diseases reported by households, people complained most about malaria (47%). Typhoid (9.6%), tuberculosis (7.2%) and diabetes (9.6%) were also reported. When we compare this with the village medical health center data reported in 2007, among the individuals who visited the hospital, 924 cases of malaria were reported, 80 cases of anemia, 46 cases of high blood pressure, 113 cases of trauma, 200 cases were reported for infectious diseases, and 129 cases for

Table 5.2. Diseases reported by households (n=83)[1] (Household Survey, 2007).

Diseases	n	Percent
Malaria	39	46.9
Typhoid	8	9.6
Tuberculosis	6	7.2
Pneumonia	5	6.0
Diabetes	8	9.6
AIDS	2	2.4
Unknown diseases	29	34.9

[1] Multiple answers.

other, non-infectious diseases. The survey research shows that about 35% of the households were not knowledgeable about the illness from which their members were suffering. Only two households mentioned HIV. In this thesis, to identify problems related to AIDS I did not focus on HIV-status (infection) but on being affected by AIDS-related illness. For this, I used health center data and observation. During the survey, households mostly did not mention their HIV status. Respondents would say they did not know the disease they were suffering from (35%). This shows the secrecy and stigmatization surrounding AIDS, and the fear of social exclusion. People know, but do not dare tell. We can assume that unknown, mostly chronic diseases, and other infectious diseases that are characterized by a loss of physical strength, are indicative that AIDS is involved.

Regarding the sources of financing the sick person's treatment, most of the affected households have got into debt, and have experienced serious financial difficulties. Among affected households, those who visited the hospital and received medical treatment (n = 48) reported different kinds and sources of financing the medical expenses. Thirty-two percent financed the treatment with savings, while 25% borrowed money from people, 13% got assistance from relatives and friends, 11% paid for the costs with cash gained from casual work, and 11% of the households sold assets to pay for the medical expenses. Table 5.3 presents the contribution to household labour in terms of the activities the sick persons were involved in.

Table 5.3 demonstrates that among the activities performed by household members, the experience of illness in the household affects crop production more (about 81%) than the other activities, i.e. poultry (6%), livestock (3.6%) and off-farm activities (7.2%).

5.3.2 Impacts of mortality on livelihoods assets and activities

In the survey, 32 households experienced the death of a household member during the past five years. Malaria (31%), AIDS (9%) and typhoid (6%) were mentioned as causes of death. Tuberculosis also caused death in some households (4%), while

Table 5.3. Labour contribution of the sick person to the household (n=83) (Household Survey, 2007).

Activities	n	Percent
Crop production	67	80.7
Poultry	3	3.6
Livestock	5	6
Off-farm activities	6	7.2
Unknown	2	2.4
Total	83	100.0

50% of the households said not to know the cause of death. The latter percentage is very high, and I tried to find out why. I suspected some kind of secrecy, related to the cause of death, which people do not want to tell the interviewers. I suspected a special apprehension about AIDS. While only two households reported AIDS as a disease experienced by the household during the past five years, nine percent reported AIDS as the cause of death of a household member. Thus the secrecy about AIDS is likely to be stronger when the HIV-positive person is alive than when he or she is dead. Nguthi (2007) has found the same.

Malaria is the most frequently mentioned cause of death. Stillwaggon (2006) has shown how malaria and AIDS are related and linked through poverty. Conditions of poverty fuel infectious diseases like malaria, which increases the prevalence of HIV and AIDS. According to her, the epidemic spreads where people are dying of other diseases that are neglected or unknown among the population. Furthermore, people who are malnourished and lack access to medical care, are vulnerable to infectious and other diseases, which increases their susceptibility to HIV and aggravates the affliction by AIDS.

It became apparent that households are faced with a labour shortage related to AIDS morbidity and mortality impacts. Results from the five FGDs show that most of the affected households have changed their land use patterns by reducing the size of land they were using in the past for food crops production such as yam, cassava, banana and maize. Accordingly, both affected and unaffected households complain about declining food yields. People look for hired labour to compensate the loss of labour. The next section gives insights on the labour shortage, and on how households and particularly women experience it and deal with it.

5.4 Labour shortage

In the survey, 67 out of the 110 households that hire agricultural labour are affected households (Table 5.4). Among the affected households that hire labour, 35 are female-headed households. This demonstrates that women are the most in need of hired labour, as they experience more pressure to reallocate the time they have available for agricultural labour to perform their domestic or gender duties. Among these duties are caring for sick people, and being available for obligatory funeral activities (Rugalema, 1999a; Müller, 2005a; Nombo, 2007; Karuhanga, 2008). Among the households who were unable to hire labour to compensate the labour loss of sick persons, 40 reported they had reallocated the work to other household members, whereas five households reported they had relatives who had moved into the household to help.

Table 5.4. The hiring of agricultural labour (Household Survey, 2007).

Hiring of agricultural labour	HIV status of households		
	Affected	Unaffected	Total
Yes	67	43	110
No	48	33	81
Total	115	76	191

Pearson chi-square HIV-status: 2.323, not significant.

The following case shows how a woman experienced the impacts of illness on her agricultural activities and food security, as well as on her expenditures, because of her husband's health condition.

Case 5.3 Shortage of agricultural labour due to a husband's illness

Ama is a resident of Y.F., aged 45. She was born in Ghana, and is married with one son, aged 18, going to secondary school. As a native, Ama's husband owns a plot of land that he inherited from his family. Ama was farming with her husband until he fell sick. She narrates the experience and impacts of her husband's illness:

'Before my husband's illness'

> *'Since six months, my husband is in this bad condition, disabled from illness. Before his illness, I woke up at 5 o'clock to perform my domestic activities, and to go farming between 8 and 9 o'clock. Our field is situated at ten kilometers from the village. I was going on foot everyday. I was farming yam, banana, tarot, pepper, tomato, aubergine, okra, while my husband was farming cocoa and coffee. We were selling our food production and were consuming it as well. On Wednesday and Sunday, the open market days, I could sell some food and get 3,000 FCFA. I used this money for cooking to provide food for the household. When I came from the field at around 4 o'clock in the afternoon, I had to go to the market to shop for food and cook for the household. I have used to go to sleep at around 21 o'clock. At that time, I spent 200 FCFA for breakfast; our lunch was made of akpessi (boiled yam and/or banana). For the diner I spent 500 FCFA for cooking foutou (pounded yam/banana/cassava). When someone from the household was sick, we relied more on traditional herbal medicine than medical prescriptions from the health center, due to the lack of money'.*

'Since my husband is ill'

> *'Since my husband is sick, I rise at 4 o'clock in the morning to take care of him, to perform activities such as preparing the traditional herbal medicine, bathing him, feeding him, and washing the laundry. After that, I go to the market to shop for something to cook for him, before leaving for farming. I can now go to farm only at 11 or 12 o'clock in the morning and come back at 14 or 15 o'clock, in time to take care of my sick husband. So, I cannot farm for longer anymore. Once I am back from the field I have to go to the market again to shop and then to cook dinner. Now I can spend only 300 FCFA on cooking for lunch and dinner. With regard to the difficulties we are facing, we cannot go beyond that, or think about food variety. Most of the time, we eat foutou with smoked fish for dinner. Every day, I have to fetch water from a well situated at 200 meters from my house, and do seven rounds to provide water for the household. Now I have difficulties fetching firewood in the surrounding fields. Sometimes, I have to go very far to get firewood. We have now abandoned the field located at 10 kilometers, since my husband cannot help me anymore with cleaning the field for crop production. Now, I do food crop production in a field located at four kilometers from here. However, the soil in this field is not fertile at all, compared to the field located at 10 kilometers. So, I cannot get the food I was getting in the other field anymore. I am still borrowing money sometimes from people in the village, to pay for medicine for my husband. Actually, he does not have any relative who can help me with any financial or non-financial assistance. I am alone, to do everything. Furthermore, during this period of illness and difficulties for us, our son was admitted to secondary school. But I cannot pay the school fees. So, in spite of all these efforts I am making, I think that the future of my son is already jeopardized' [Ama, Y.F., 28-02-08].*

Case 5.4 Re-allocating agricultural labour to take care of a sick sister and son

Florence is a resident of Y.F. village. She is aged 46, and was born in the village. She is married and has studied up to secondary school. Florence has five children and three sisters. Some of them live in the village, whereas some are in Abidjan, the economic capital. She has to look after her children along with her sisters, as she is the eldest. When they are sick, she is always the one who is supposed to take care of them. Below, she narrates part of her experience with the re-allocation of agricultural labour due to illness, and the related impacts on her household's food and livelihood security.

> *'On the 8th of September 2007, my son had an accident. He was admitted to the hospital for three weeks. I was there alone, taking care of him, without his father. The expenses were estimated at 615,305 FCFA. I still wonder*

> *how I can refund this money I borrowed from people. The same year, my sister fell sick. At 5:30 I had to wake up to start doing the domestic tasks, bathing my sister, washing her laundry, cooking her food, preparing the medicines she has to get, and accompanying her to hospital when she had an appointment. So, during one full month, I could not go to farm. Not only did I have nobody who could go to farm for me, but I had no money to hire people to perform my agricultural work either. However, even though I could not attend different meetings in the village because of my sister's illness, I had to participate in the funerals. When there is a death, everyday, women have to provide some help for cooking, to take care of the people gathered for the funerals. During funerals, you are not allowed to farm on Friday and Saturday. Friday is for the watch, and Saturday is for the burying. Finally, our agricultural calendar dropped from six days to three days, often less than this, because of the funerals. Currently, there are funerals almost every week. If it is not in the village, it is in villages around. So, you have only two or three days in the week to perform your agricultural work, on Monday, Tuesday and Thursday. If the funerals are from your close family, you should observe one month without going to farm. In fact, as long as the guests do not leave, you have to take care of them' [Florence, Y.F., 28-02-08].*

Women experience more trouble in the allocation of their time to agricultural labour because of the obligation to participate in funerals, but also with cases of illness in their household and extended family. In this regard, they need to find a replacement within their kinship network, or hire labour for crop production to feed their households. Agricultural wage workers, called *domake,* who can be both Agni and migrants, work individually or in groups, depending on the agricultural tasks to be performed.

During the past five years, the price of hired labour[17] has increased, due to the war in Côte d'Ivoire and its impacts on the economy, and this is also visible in the soaring of food prices. Because of the war in 2002 and its aftermath, many migrant wage workers moved back to their home countries, like Mali, Burkina, and Ghana, which had an impact on the availability of wage workers in rural Côte d'Ivoire. Thus, it has become more difficult for women to afford hired labour. From participant observation and in-depth interviews, it transpired that AIDS-related socio-economic hardship may increase some women's exposure to transactional sex with the agricultural workers, as a coping strategy to afford hired labour.

Furthermore, results demonstrate that in conditions of economic hardship, some people may change their work to cope with food and livelihood insecurity. Below, a case portrays this fact.

---

[17] The price of hired labour has increased from 65,000 FCFA to 120,000 FCFA, for a contract of one year. The price is relative to the scale of the agricultural labour to be performed. Yet, some people may negotiate.

Case 5.5 Poverty and food insecurity changes people's work

Rama is a migrant woman coming from Mali. She is married with seven children. Like most of the migrant households, she lives with her brother-in-law and his family, i.e. his wife and his eight children. When she got married and joined the family, the livelihood was based exclusively on the trading of smoked fish, imported regularly from Mali. It was the household business, managed and lead by her late father-in-law, her brother-in-law, and her husband. In the past ten years, the household livelihood has changed. They are now farming maize, as an alternative that enables them to feed the household. The business went bankrupt, since people no longer have money to buy their smoked fish, like they did in the past. As she stated, she never realized that the household would one day stop the smoked fish trade and go into agricultural activities. Rama has this to say:

> *'Agricultural activities are not our work or business. But we were obliged to do it, because the trading is not working out anymore. We have no other way to feed the household. So, we are now doing agricultural activities to earn the household consumption' [Rama, Y.F., 28-02-08].*

AIDS-related morbidity and mortality are seriously affecting household food and livelihood security. Besides the hardship, time and cost-related funerals also add to people's livelihood and food insecurity. In Agni culture, more than among the migrants, funerals are not only costly, but also time-consuming. There is a cultural obligation to attend. I could witness how funerals impact on Agni households and jeopardize their food and livelihood security. In the next section, I give a picture of funerals in the Agni community and the socio-economic consequences they have for Agni households.

5.4.1 Funerals in the Agni community and their economic impacts

It has emerged from the five FGDs that, for the past ten years, the village has been suffering from relatively more deaths, mostly of young people and adults, that is, of active people. In the Agni community, funerals are not only costly, but also time-consuming. They impact seriously on agricultural labour time. Eschlimann (1985)[18] shows the importance that funerals have in the Agni community in Côte d'Ivoire. For the Agni, funerals are the most important event, together with *la fête de l'igname*. For this reason, the whole community is expected to participate. According to Eschlimann

---

[18] 'Participation at funerals becomes an obligation that can not be escaped, especially if the person has a kinship relation with the deceased. One abstains from agricultural work even if it suffers from the absence; but being absent from the funeral gathering is worse. The money to pay for the care of the illness may not be there, but one takes on debts at whatever price to comply with the funeral obligations' (Eschlimann 1985:11).

(1985), the cost-related funerals impact very much on Agni social life. Many among the Agni people are usually faced with debts because of the costly funerals.

Among the Agni, there are two kinds of funeral practices: when the deceased is the first, second or third child of a family, the funerals have to be short. These funeral practices are called *efewah*. However, from the fourth child onward, the funeral practices imposed by the culture have to be well-organized and last longer. These are called *essengh*. These funerals are known to be not only time-consuming, but also very costly. Three days are taken for the rituals, in which people are obliged to assist. They are not allowed to farm during this period. In the research area, people complain about the expensive funerals. The narrative below gives some insights on the cost of Agni funerals in this area.

> *'When you lose someone, you should pay for some cloths for the burying, among which clothes is one 'Bomoh', an expensive one. For the watch, you should give a contribution for food and drinks. The contribution ranges from 20,000 FCFA to 30,000 FCFA. The biggest stress we have in this village when organizing funerals is about food. In this village, compared to the past, food is the most expensive item, and one of the reasons why funerals are said to be costly in Agni culture. This may be a paradox when one lives in rural setting, but this confirms the decrease in food productivity we are experiencing in this place. Commonly, in Agni culture, funeral expenses range from 400,000 CFA to 500,000 CFA for the poor and from 1 million to 5 million FCFA for rich people living in towns. In the village, people have an obligation to make a financial contribution, from 200 FCFA to 500 FCFA, to assist the family plunged into mourning. According to Agni culture, even children have to provide for the contribution. They pay 200 FCFA, whereas adults pay 500 FCFA. So, it is mothers who are paying for their children' [Akoua, Y.F., 28-02-08].*

Illness and funerals add to people's vulnerability, especially to that of women. For them, the direct impact of illness and funerals is the loss of labour, as they have to re-allocate their agricultural labour to caring for the sick and to participating in funerals. In the case of migrants, however, funerals are not so costly and time-consuming. Muslim culture demands that funerals are organized within one day. Both Agni and migrants assist each other during funerals in their respective communities. However, the contribution by migrants to Agni funerals is less than it is *vice versa*. So, Agni households experience more socio-economic impacts of funerals than migrant households do.

In conclusion, the results shed light on the social and economic impacts of illness and death on household assets, agricultural activities and food security. AIDS has a direct effect on agricultural production and food security through a reduction of women's time allocated to farming (cf. Rugalema, 1999b; Müller, 2005a; Nombo, 2007; Karuhanga, 2008; Nguthi, 2007).

5.4.2 Social capital and internal vulnerability

For both the Agni and migrant communities, social capital is an important asset that is culturally enrooted. Households may rely on their participation in organized groups as an important asset to cope with the food and livelihood insecurity they are faced with. There exist not only formally organized groups, but also groups of people who get together regularly to do an activity. These local organizations are mostly farmer groups or cooperatives, neighbourhood committees, and savings groups.

It has been difficult to make a clear distinction between formal and informal social networks, as they are all shaped by social relations (Nombo, 2007). The local organizations here are small-scale and not legally registered. Some farmers are organized in groups to provide their service to big farmers looking for hired labour. These organized farmers' groups exist among both Agni and migrants. A group of Agni women have organized themselves in a cooperative for food crop production, to meet and work together in order to challenge the hardship they are going through. Saving-and-credit groups exist in both communities. Their members are mostly women who have a limited financial capacity. Each group member has to daily contribute 100 FCFA. Daily, the total amount of the contributions, which is not more than 2,000 FCFA, is given to a member of this informal saving-and-credit network.[19] Besides the importance of access to formal credits and financial services, Bekele (2010) shows the importance of the participation in credit groups for women's empowerment: 'Extending credit to women promotes women's opportunity for self-employment activities. Apart from financial services, participation in credit groups is considered to be a powerful tool in raising women's self-confidence. It also serves as an entry point to advance women's social and political empowerment' (Bekele, 2010: 2).

Participation in religious groups and burial groups was also reported. In terms of the major religious groups, Roman Catholics and Muslims have groups of households who meet and pray together and share spiritual guidance. For the Agni, the burial group consists of Agni households and Agni individuals. For migrant Muslims, the religious group also performs as burial group. Non-Muslim migrants may rely on their religious groups, Roman Catholic or Protestant, as a burial group. In these burial groups, members contribute both services and money to the group. The group provides support and assistance to members in the event of death (Nombo, 2007; Nombo and Niehof, 2008). However, results from Agni men and women FGDS as well as from interviews with Agni leaders demonstrate that the financial contribution

[19] For the informal and saving network, women are organized in small groups of 20 people. Five informal and saving networks have been reported, for both Agni and migrants.

to the Agni burial group is a burden for Agni households. The contribution is fixed, and non-negotiable. Each adult should provide 500 FCFA, and pay 200 FCFA for each child in the household. For the migrant, the financial contribution is not strict and not fixed either. Children do not count as they do in the Agni community. The membership size for these local organizations ranges from 10 to 100, except for the Agni burial group, which includes many Agni households and individuals living both in and outside the village, as they belong to the same Agni community.

People participate in such local organizations as they may look for support from the group to cope with hardship and crisis. In the survey, respondents were asked about any group membership of themselves and other household members. Over 30% of the households mentioned they have a member involved in such networks. Among these households, 33% complained about the lack of time due to illness and the lack of financial means, which hamper them to attend meetings, and to contribute the obligatory financial contribution. In this way, AIDS-related economic hardship impacts on social capital, as vulnerable households, particularly women, are unable to maintain their participation in social networks, due to the lack of financial resources and time to invest and be active in the groups. Consequently, this deprives some households of any consistent support they may gain from their social capital to cope with economic hardship, as social capital at the community level can no longer support vulnerable people (Nombo, 2007; Nombo and Niehof, 2008; Karuhanga, 2008; Nguthi, 2007). In such a context, it is difficult for people to make ends meet for their households' food security.

5.5 Food security, nutrition adequacy

This section presents findings from the quantitative data on households' food insecurity, using indicators such as the food crop production at household level, food shortages experienced by households during the past two years, the quantity of food stored by a household, the number of meals per day for household members, and the composition of the household menu in terms of protein and food commodities. Additionally, the section presents data on household food insecurity, based on experiences and perceptions obtained from FGDs with men and women from both Agni and migrant clusters, as well as on individual interviews with local leaders.

In terms of food crop production by the household, yam is the most produced crop, followed by cassava, banana and maize (Table 5.5). This reflects actual food habits in both Agni and migrant clusters. For the Agni, the basic local staple food is *foutou*, a mix of pounded banana, yam or cassava, which people eat with different types of sauce; and *akpessi*, known as boiled or smoked cassava, yam and banana. For migrant people, *kabato, kokohndehn* and *malo* are the basic staple foods. *Kabato* is made of maize flower, *kokohndehn* is made of cassava flower, whereas *malo* is cooked rice. These staples are eaten with different types of sauce as well.

Table 5.5. Food crop production at household level (n=191) (Household Survey, 2007).

Crop production	n[1]	Percent
Yam	51	26.7
Maize	16	8.3
Cassava	32	16.7
Taro	11	5.7
Banana	21	10.9
Rice	5	2.6
Egg plants	19	9.9
Tomato	6	3.4
Pepper	9	4.7
Okra	6	3.4
Courgettes	25	13.0

[1] Multiple answers, hence n>191.

Above, we defined food security as 'access by households at all times to enough food for an active and healthy life of the household members' (Niehof, 1999: 107). Based on this definition, the survey results demonstrate that many households experience food insecurity, which was confirmed by the FGDs. Table 5.6 gives insights on the incidence of experienced food shortage during the past two years, according to the ethnic status and HIV status of the household.

Table 5.6 shows that Agni households have experienced significantly more food shortages during the past two years than migrant households have. The association with HIV status is also significant. According to the HIV status of households, the majority of the households (63 households out of 191) that experienced food shortage during the past two years are affected households.[20] Only 27 unaffected households reported that they experienced food shortage within this period.

In order of priority, foods stored are yam, cassava, banana and maize. In terms of the quantity of food storage according to the HIV status of households, it has emerged that 95 affected households stored food, against 69 unaffected households (Table 5.7). But the quantity of food that households reported to have stored is not that much. The quantity of food in storage is more often about 1-5 sacs than 6-10 sacs. According to ethnicity, results show that Agni households store food more than migrant households do (85 against 79 households). It has been noted that migrants

[20] See Chapter 4 for the operational definition of affected households.

Table 5.6. Incidence of experiencing food shortage during the past two years according to the ethnic status and HIV status of households (Household Survey, 2007).

	Experienced food shortage		Total
	Yes	No	
Ethnic status of households[1]			
Agni	62	40	102
Migrant	28	61	89
HIV status of households[2]			
Affected	63	52	115
Unaffected	27	49	76
Total	90	101	191

[1] Pearson chi-square ethnic status: 18.10, $P<0.01$.
[2] Pearson chi-square HIV status: 6.81, $P<0.01$.

are more in the habit of buying food than the Agni are, as they are landless. With the options of renting land for food production and sharecropping, they may produce more food, but not necessarily enough to sustain them.

Table 5.7. Quantity of food storage during the past two years according to the ethnic status and HIV status of households (Household Survey, 2007).

	Quantity of food storage (in sac)		Total
	1-5	6-10	
Ethnic status of households[1]			
Agni	85	17	102
Migrant	79	10	89
HIV status of households[2]			
Affected	95	20	115
Unaffected	69	7	76
Total	164	27	191

[1] Pearson chi-square ethnic status: 3.304, not significant.
[2] Pearson chi-square HIV status: 2.523, not significant.

The results analyzed according to the HIV status of households and according to ethnicity demonstrate that most people have three daily meals (Table 5.8). There is no significant association with HIV status, but there is with ethnicity. The proportion of migrant households having three meals a day is higher than that of the Agni households.

According to Table 5.9, HIV-affected households eat more snails (protein) than the unaffected households do. Agni households also reported to consume more snails (54 households) than meat and fish, whereas 39 migrant households have reported to consume fish, 24 households eat meat, and 22 households eat snails as part of their daily meal. The chi-square for association with ethnicity is highly significant. In fact, sauce made of snails is eaten in combination with *foutou* of yam, cassava or banana, which is the basic staple of the Agni. Snails are gathered from the surrounding forest and serve as a no-cost alternative for fish. Moreover, there are less men hunting, and the forest's ecosystem depletion reduces the presence of game.

Both Agni households and HIV-affected households have experienced a significant food shortage during the past two years (see Table 5.6). However, Agni households are the most HIV-affected (55%), compared to the migrant households (45%), so that could explain the effect. Yet, from FGDs with Agni and migrant women as well as from participant observation, I came to realize that, because of pride, both the Agni and migrant communities hide the truth about their households' food insecurity. People do not want outsiders to know how food insecure they are, because this is seen as somewhat shameful for the household. Nombo (2007) found the same in

Table 5.8. Number of meals per day according to the ethnic status and HIV status of households (Household Survey, 2007).

	Number of meals per day		Total
	≤2	≥3	
Ethnic status of households[1]			
Agni	39	63	102
Migrant	15	74	89
HIV status of households[2]			
Affected	34	81	115
Unaffected	20	56	76
Total	54	137	191

[1] Pearson chi-square ethnic status: 14.810, $P<0.05$.

[2] Pearson chi-square HIV status: 1.055, not significant.

Table 5.9. Consumption of protein according to the ethnic status and HIV status of households (Household Survey, 2007).

	Consumption of protein				Total
	meat	fish	snail	meat+fish+snails	
Ethnic status of households[1]					
Agni	9	22	54	17	102
Migrant	24	39	22	4	89
HIV status of households[2]					
Affected	23	36	42	14	115
Unaffected	10	25	34	7	76
Total	33	61	76	21	191

[1] Pearson chi-square ethnic status: 43.577, $P<0.01$.

[2] Pearson chi-square HIV status: 3.443, not significant.

her study. Moreover, I came across biases in the perception, interpretation and definition of what constitutes a daily meal in both communities. What people call a daily meal is anything you can eat during mealtime, irrespective of the quantity and quality of such a meal. I could observe people eating two or even one time per day in this setting, without having food such as egg, milk, bread, or sugar. In Agni households, children and adults use to eat *akpessi* (boiled or smoked yams, banana or cassava) for breakfast and lunch, and *foutou* (pounded banana, cassava or yam) for dinner; whereas in migrants households, they use to have *baka*, (a porridge made of maize flower) for breakfast, *akpessi* for lunch, and *kabato* or *kokohndehn* (made of maize and cassava flower) for dinner. The soups accompanying these daily meals are mostly made without meat. Indeed, during the FGDs, both Agni and migrant women acknowledged that they rarely eat meat:

> *'From your husband, you can get only 200 FCFA and often nothing, for the daily cooking. You as a woman are then obliged to find money so that you can cook for him and the children. To be honest, we can state that we eat once a day: children have a smoked banana for breakfast and lunch. You, as the mother, you are often obliged to fast so that the children can eat something. We women have hunger for a habit. Actually, we are very malnourished here. Food commodities exist in this village. But due to the lack of money, we cannot afford them. Our children are always sick. They are suffering mostly from anemia and malaria' [Flora, Y.F., 27-04-07].*

> *'We get 200 FCFA (25 Eurocent) from our husbands for the daily cooking. What can you buy for 200 FCFA? You are then obliged to buy dry fish powder for 100 FCFA, okra powder for 25 FCFA; onions for 25 FCFA, tomatoes for 25 FCFA, and spices for 25 FCFA. This is how and what we are eating here. Our children are sick most of the time, suffering from malaria, anemia and dysentery' [Ami, Y.F., 27-04-07].*

To summarize, both Agni and migrant households are experiencing problems of food security. This is confirmed by the survey data on food shortage experienced by households and on households' consumption of protein. What is more, even though Agni women have a more secure access to land (see Figure 5.1), their households are less food secure. Moreover, the survey results are likely to underestimate the actual level of food insecurity, because people are embarrassed to report food shortages.

5.5.1 Changes in the daily menu and food coping strategies

This research shows that 38% of the affected households and 25% of the unaffected households have changed their daily menu and have introduced new foods to deal with the food shortages. During the FGDs, Agni women stated that they are now eating *kokohndeh*, *kabato* and *malo*, which they were not eating in former times. The migrant women acknowledged that, even though they have introduced *akpessi* of banana, yam and cassava into their food habits, they rely more on *malossainin* and *kabagnignin* as food coping strategies during food shortage periods in May, June, July, and August. Thus, Agni women as well as migrant women have developed some food coping strategies to deal with the seasonal food shortage. Below, an Agni woman tells how Agni women cope with food shortages:

> *'Normally, we are supposed to eat tree times a day. We did in the past. But for some years now, we cannot, because of the food shortage. This situation has motivated us to find ways to overcome our hunger during the dry season. We have become dioula women (migrant women), because we now eat their basic staple food, I mean kabato and kokohndehn. We even have a recipe of kabato made with dry banana, to deal with the seasonal food shortage in the village. So, the problem of food we are facing here allowed us to get many new recipes to overcome the hunger. We can call them the recipes of the food crisis' [Mado, Y.F., 27-04-07].*

Agni women now have 'food crisis recipes', as they call it. *Kabato* and *kokohndehn* have become their food coping strategies, because during times of food shortage, banana and yam become so expensive that many households can no longer afford to enjoy their preferred staple, i.e. *foutou*. Thus it is easier for them to cook *kabato* and *kokohndehn*, which is made respectively from maize and cassava flower. They mentioned that they prefer these foods over the other types of foods during times of food shortage, because maize and cassava are relatively cheaper than banana and yam.

For migrant women, using *kabagningni,* and *malossainin* is a way to deal with food shortages. These staples are made of the bottom food crust of cooked rice (*malossainin*) and maize couscous (*kabagningnin*). After cooking rice (*malo*), women keep the bottom food crust, and recook it, mostly with only some oil and salt added. The maize couscous is recooked as well, with nothing but some oil and salt. In short, migrants mentioned that they prefer these foods over other types of foods, because relatively, these are the cheapest foods they can afford, particularly during times of food shortage. Below, a migrant woman tells how migrant women cope with food shortages:

> *'Actually, we cannot afford eating like we could in the past. Due to the lack of money and the size of the family, we developed some mechanisms to deal with periods of food shortage. For instance, during these periods, we eat more kabagningnin, a couscous of maize cooked with only some oil and salt. We do the same thing with malossainin' [Sali, Y.F., 27-04-7].*

5.5.2 Socio-cultural food practices and household food distribution

FGDs with both Agni and migrant women indicated that the food distribution in households is increasingly unequal, since the head of the household is supposed to get the biggest part. Women acknowledge that they are always last.

> *'About food distribution: the head of the household is supposed to get the biggest part. You cannot refuse to do so; otherwise, he can even beat you for that. Most of the times, children don't get fish or meat. The head of the household is the head of the food' [Cathi, Y.F., 27-04-7].*

Thus, the intra-household food distribution is related to the decision-making processes within the household. It appears that when the head of the household is male, he may have a preferential access to food, at the expense of children and women. This makes it difficult to efficiently divide the food available in equal portions amongst the different household members. Food distribution in favour of the male head of the household lies more with a survival strategy, adopted by women to protect themselves from getting beaten, and to escape the potential threat of getting divorced. An unequal food distribution may be a result of the male head of the household's decision-making power (FAO, 1999).

Besides, results from the five FGDs and interviews with local leaders show that households observe food taboos. Some people do not eat goat, snake, antelope meat, and some varieties of fish. Despite food shortages, food taboos can still be observed. Remarkably, despite their economic and food vulnerability, households continue to engage in food exchange practices. Survey data show that almost 30% of the affected households and 47% of the unaffected households have reported the exchange of food, particularly with neighbours and the husband's relatives, especially during ceremonies. In the FGDs with Agni and migrant women, women stated that they cannot stop this practice, because it is historically and culturally rooted in both communities.

5.6 Conclusion and discussion

Both survey data and cases studies demonstrate that the morbidity and mortality related to AIDS are affecting households' food and livelihoods security. Using factors such as the education and gender of the household head, the marital status of the household head, access to land, access to hired labour, as indicators of internal vulnerability, the results show that labour shortage adds to the factors affecting household food security. It shows that, compared to their male counterparts, women in an AIDS context experience more agricultural labour shortages because of their gender duties, among which is the duty to reallocate time to care for the sick and to take part in funerals in the neighbourhood. These findings of a West African country case study comply with findings from East Africa (Müller, 2004, 2005; Rugalema, 1999a,b; Nombo, 2007; Karahuanga, 2008). The study shows that Agni as well as migrant women experience vulnerability, but in different ways. Although Agni women have access to land, they are food insecure because they experience more labour shortages, due to the importance and burden of funerals in their culture. On the other hand, migrant women have no access to land, but they can buy food on the market from money they earn by working on Agni land and by running small, off-farm activities.

In Case 5.1, Cathi got land use rights from her late mother, although land ownership and the use right from inheritance is not culturally acknowledged for Agni women. In Cases 5.3 and 5.4, Ama and Florence showed that access to land is not so much a problem for Agni women as the lack of (male) labour within the family due to illness is, and the lack of time to farm, due to the time that must be allocated to caring for the sick and attending funerals, along with the related financial impacts. The combination of these conditions severely adds to their internal vulnerability, as it impacts on their food and livelihood security. In Case 5.2, Sali showed how landlessness adds to the vulnerability of migrant women. As a female migrant, she does not have access to land from the autochthonous Agni. However, the migrants' own patrilineal land tenure and inheritance system does not allow her to get access to land either, otherwise than through her husband or male relatives (Spring, 2000). As she is lacking money to rent land, the alternative of sharecropping is also unsuitable to guarantee her household food security, because of her household's large size. Case 5.2 portrayed how poverty and food insecurity force migrant women to engage in agricultural activities that they are not used to, in order to access food to survive.

Households may rely on their social capital to cope with the internal vulnerability. Its transpires that AIDS-related economic hardship impacts on social capital, as vulnerable households face constraints to maintain their participation in social networks, because they lack the resources and time to invest in and sustain their social capital (Nombo, 2007; Nombo and Niehof, 2008). Still, both Agni and migrant affected households continue to engage in food exchange practices.

This study shows that matrilineal and patrilineal kinship provide different potentialities and constraints for Agni and migrant women, respectively. For example, among the Agni women, some may have several partners to ensure that they have children and can perform their culturally prescribed reproductive role. Patrilineal migrant women, on the other hand, have to be submissive to their husband's sexual desires, even in the case of AIDS, at the risk of becoming stigmatized in their community for not complying with norms about the behaviour of married women.

This study shows that AIDS is not a direct causality in people's livelihoods and food insecurity, but that it does aggravate livelihood vulnerability and food insecurity. From a gender perspective, it shows that as long as poor women are overburdened with agricultural work for food production, and with the care for sick persons, they are more likely to be exposed to AIDS than their male partners are (Nombo, 2007; Nguthi, 2007; Karuhanga, 2008). In terms of ethnicity, the differences and similarities observed between Agni and migrant households are caused by different factors, including AIDS, household size, access to land, and the overall internal vulnerability. This means that vulnerability is the outcome of an interaction of different factors and processes, which are experienced differently by Agni and migrant households and individuals within the household. Moreover, the respective positions of Agni and migrant women in a matrilineal or patrilineal kinship system produce qualitative differences between the factors and processes that contribute to their individual and household vulnerabilities. This study shows that AIDS is not only a medical problem, but a socio-cultural problem as well. Socio-cultural features, such as matrilineal and patrilineal kinship systems, may potentially increase women's exposure to the disease and increase vulnerability to its impacts. Even though women cannot escape from their culture, they may be capable to negotiate and challenge some of the norms and practices of matrilineal and patrilineal systems, albeit within limits. Therefore, AIDS mitigation policies and interventions should better address the social and cultural differences at the micro level that may explain the variation in women's vulnerability to AIDS.

Chapter 6.

External vulnerability of Agni and migrant households

This chapter discusses the external vulnerability of Agni and migrant women and their households in relation to the ecological, institutional, and economic context. The material for this chapter is derived from the five Focus Group Discussions, interviews with Agni and migrant women and with local leaders, key informants, as well as from participant observation. The first section addresses the external dimensions of food security in terms of crop production and women's access to extension services, credit and markets. The second part of the chapter discusses their vulnerability due to ecological degradation. The chapter continues with the presentation of data relating to the health situation in general, and the AIDS situation in particular. I will discuss malnutrition in this section as well. The next section deals with the relationship between social capital and external vulnerability. The chapter ends with a general discussion.

6.1 Food and livelihood security, access to agricultural extension services, credit, and markets

6.1.1 Food crop production

Although the Abengourou region has been important for the cocoa production, in this study, the focus is on the understudied food crop production. The main food crops cultivated in the village are yam, cassava, banana, and maize. Within one year, food productivity has declined about 30%. According to the extension officer, ANADER data indicate that food production in Y.F. dramatically dropped in 2006 (see Table 6.1).

There seem to be two causes for the decline: the number of households that are engaged in food farming has been dramatically reduced, and women's cultivation of tomato, okra, spanish pepper and eggplant has plummeted. These are the vegetables that need regular attention, something the women no longer have time for. They seem to have replaced these vegetables by the less time-consuming cultivation of courgettes. Secondly, the acreage has been reduced or is cultivated less intensively, as is the case with maize. Cassava, on the other hand, is planted more frequently. The acreage planted with yam, being the favorite staple of the Agni, remains remarkably similar. However, since the heavy duty of making the mounds (*buttes*) for the yams is a man's job, it is recorded that now even women take up this heavy work, or they hire labourers for it.

Table 6.1. Y.F. food crops production in 2005 and 2006 (Source: ANADER, 2007).

Crops	2005				2006			
	Number of farmers	Area (acres)	Yield (tons/ha)	Production (tons)	Number of farmers	Area (ha)	Yield (tons/ha)	Production (tons)
Banana	561	463	10	4,630	84	84	5	420
Cassava	851	393	15	5,895	706	476	10	4,760
Yam	844	709	16	11,344	852	787	12	9,444
Maize	611	557	1.5	835.5	404	172	1	172
Rice	58	56	0.8	44.8	71	47	0.5	23.5
Okra	276	83	2	166	64	43	1	43
Spanish pepper	233	62	2.5	155	127	40	2	80
Tomato	238	137	8	1,096	93	14	5	70
Eggplant	219	74	6	444	90	42	5	210
Courgette	120	75	13	975	144	98	10	980

The decline in food productivity observed in the area obviously influences the availability of food, making it difficult for households to get access to sufficient quantities of food and appropriate foods for a nutritious diet (see Chapter 5). In addition, food insecurity experienced by households is mainly due to poor access to extension services and credit, which are needed to improve the situation.

6.1.2 Access to extension services

Although agricultural labour is the main livelihood activity in the village, small businesses, such as businesses for food and drink processing and selling it locally, are important as well. From the survey research, it has become apparent that the majority of households possess traditional farming knowledge and skills. Farmers lack the financial means to modernize and improve their agricultural production, even if they want to. This pictures a lack of access to the technology and extension services needed for an agricultural production that guarantees food security. Observation shows that it is hard for people to get access to and afford agricultural extension services, because it is costly for them. In fact, while officially, the National Institute for Rural Development (ANADER) should provide free extension services to farmers, as claimed by the ANADER officers during interviews, in practice, the farmers state that they have to pay for these services. The lack of access to agricultural services is a proxy of external vulnerability, as this increases rural people's and particularly women's livelihood insecurity, and therefore affects food productivity.

Formally, ANADER has one extension worker on food crops at village level, and 43 at district level. From the FGDs, however, as well as from the interviews with local leaders and the ANADER officers themselves, a different picture emerges about the agricultural extension services' performance. As the village extension officer states:

> *'The extension services from ANADER are free of charge for farmers. Let me take a specific example. When a group of women farmers ask for our services, we do a need assessment and provide the technical assistance they are looking for. After that, we do the follow-up. However, farmers themselves should supply seeds and fertilizers. We notice that, compared to men, women cannot afford that. Actually, this is the major reason of the difficulties they have to improve their food production, we acknowledge that' [ANADER extension officer at Y.F., 25-04-07].*

In contrast to this statement, in FGDs with both Agni and migrant men and women, people stated:

> *'When you ask for the services from an ANADER extension worker, not only do you have to pay for the fuel to get him to your field, but you also have to pay for the seeds and fertilizers. Where can we get such money? [Madeleine, Y.F., 27-04-07].*

These different perspectives on women's experiences and the ANADER narrative about its own performance reflect the lack of government support to agricultural services and the low provision of inputs to mitigate rural people's food and livelihood insecurity. Given the magnitude of rural women's impoverishment and their key role in food security, the national agricultural policy should make subsidies available to provide for the seeds and agricultural technical assistance the women need to sustain their livelihood and food security.

6.1.3 Access to credit

The issue of the lack of access to formal credit for women was addressed during FGDs with Agni and migrant women. I had observed that there is a microfinance institution operating in the village, but could not understand why women in Y.F. did not have access to credit. The manager of the microfinance institution, which is called CMEC, explained to me the objective and expectations of the institution, and the way the institution is performing. The *Caisses Mutuelles d'Epargne et de Credit* (CMEC) is one of the private microfinance institutions operating in rural areas in Côte d'Ivoire. According to the manager, CMEC seeks to make credit available to rural people, to finance their personal or group projects. To be a CMEC client, one has to pay an entrance fee of 5,000 FCFA and a contribution of 3,500 FCFA[21], with a minimum initial deposit of 1,500 FCFA. Besides, when the client fulfills the admission's requirements, he or she is supposed to get at least three months of savings, with a physical asset as a guarantee, to get access to a credit of between 15,000 FCFA and 100,000 FCFA. The interest rate is about 18%, whereas the monthly charges are 250 FCFA. Even if, like in the case of Agni women, they have the right to inherit land, in many cases, they are not allowed by their father or brother to use it as collateral. Moreover, from their harvests, women can only get an average cash income ranging from 10,000 FCFA to 5,000 FCFA. This is not sufficient to obtain a CMEC credit.

This is why most women do not have access to credit. CMEC fees and interest rates are beyond their financial capabilities. Therefore, CMEC is far from achieving its goals, i.e. to alleviate poverty and food insecurity in rural areas, unless the institution becomes more sensitive to rural women's socio-economic problems.

From the FGDs with migrant women, I learned that some women living in the village were not even informed about the existence of the CMEC. Those who were informed said that the CMEC is not made for them, since they are far from being capable of meeting the required conditions to get access to credit from the institution.

> *'Without money, how can we save to get credit from the CMEC? This means that the CMEC's credit is not created for us' [Ami, Y.F., 27-04-07].*

However, the CMEC manager acknowledged the limitations of the microfinance institution in having an impact on women's socio-economic situation in the area:

> *'At the beginning, the CMEC's target were rural women. The overall aim was to help them improve their livelihoods activities with small credits to overcome their poverty. Unfortunately, we realized that men got access to CMEC credits more than women did. Why? Because women said they are incapable of responding to the CMEC's requirements for getting access to credit. The membership fee, which is about 10,000 FCFA cash, is actually too much for them. They cannot afford it. Besides, we noticed that the rare women who got the credit used the money, not to develop their livelihood activities, but to resolve their husbands' and children's problems. With regard to the extreme socio-economic vulnerability women are experiencing in this village, I could understand that they react in terms of urgency to resolve their immediate problems. Furthermore, we have 18 men and only 3 women in the executive committee, because to be a member of the board you should already have an*

[21] One Euro was equivalent to an average of FCFA 665 at the time of the field research.

> *account at the CMEC. Since women have problems to become members, we could not get many women onto the board. Therefore, I acknowledge that the CMEC has missed its goal, to improve women's socio-economic position in this village. Actually, we have to rethink the CMEC policy, if we really want to help women here' [CMEC manager, Y.F., 27-04-07].*

Microfinance institutions such as the CMEC, operating in rural areas, are not functioning as they intended with regard to targeting rural women. Not only are women unable to afford the membership cost, but also the interest rate is too high to impact on women's socio-economic situation. In its overall operation, not only is the CMEC lacking visibility and failing to target the women in the village, but it does not provide a follow-up for allocated credits either, or training on credit management skills for beneficiaries. It fails to realize the institution's goal, that is, to alleviate poverty.

Results show that, even though women do not have access to formal institutions' credits, they do have access to savings and credits from relatives and other connections (see Chapter 5), to deal with sudden expenses such as in cases of illness and funerals. Besides, through participant observation, I came across Agni and migrant women having savings-and-credit exchanges among themselves, although I realized that such exchanges do not have much impact on their livelihoods security.

In other words, the lack of access to formal (external) credit institutions is also an indicator of external vulnerability, because it reduces women's ability to safeguard their household's food and livelihood security. Credit from formal institutions could possibly help women, to improve their on-farm and off-farm activities, provided the CMEC creates better criteria for women to access credit, like group membership and collective collateral.

6.1.4 Access to markets

I witnessed the difficulty both Agni and migrant women are faced with when it comes to selling their produce. During FGDs, Agni women remembered that they had to resign from a cooperative they had created for crop production, because of the lack of resources and means to get access to the markets in Abengourou and Abidjan (see Chapter 7). The cost of the transportation was so high[22], that it influenced their business activities considerably. They remembered that they could not make any profit out of it. Therefore, when they reorganized for a second cooperative, they definitely did not want to engage in selling crops, although this would give an

[22] To be able to sell their food crop products, Agni women had to rent a truck at 30,000 FCFA to get access to the markets in Abengourou, the district's main town; and 100,000 FCFA to get access to the markets in Abidjan, the economic capital and business center.

economic added value for their livelihoods. For this reason, they opted to sell their produce to the retail sellers, among whom are many migrant women.

Being landless, migrant women may rely more on the business of food crops reselling. They buy food products such as cassava, banana, yams, vegetables and fruits from those who farm and make a profit by selling these to people in the surrounding areas. Nevertheless, I realized that this business does not have much impact on their socio-economic condition. I found out that one of the main bottlenecks is the cost of transportation, which did not allow these women to get a consistent profit out of it.

In conclusion, for both categories of women, the lack of institutional and physical infrastructure and transportation facilities, which are the indicators of structural poverty, reinforces women's livelihood insecurity, and jeopardizes household food security.

6.2 Environmental degradation as external vulnerability

Apart from the problems related to poor access to extension services and credit that contribute to the decrease in food production, people brought forward forest degradation as being part of the vulnerability context and a cause of crop failure. I learned from in-depth interviews what was defined and perceived as 'forest degradation' by local people and by forestry officers, and how women experienced its impacts on livelihood and food security. In this study, forest degradation is treated as external vulnerability, since the lack of good soils for agricultural production, due to the exploitation of natural resources, jeopardizes rural people's livelihoods and potential development. What is more, while households are faced with fragile soil conditions due to the ecological instability that is caused by a lack of rainfall, they do not have the resources and means to invest in natural resource conservation. Yet, they are also faced with the lack of investments in conservation techniques from the government, to challenge the effects of forest degradation. Field observation shows that the lack of good soils affects women more than their male counterparts. For this reason, in this thesis we see forest degradation as an external vulnerability factor that impacts negatively on household food and livelihood insecurity. It may potentially increase women's exposure to AIDS, when they come to deal with the hardship and poverty by engaging in transactional sex for food or cash, in order to be able to feed their households.

6.2.1 Narratives of forest degradation

The forestry official's narrative

Forest degradation is usually defined as the complete clearance of tree formations (closed or open) and their replacement by non-forest land uses. Forest degradation refers to changes in crown cover as the combined result of forest clearing and degradation, and is then defined as: 'a human-induced or natural reduction in – or even the complete removal of – tree crown cover in a given natural forest area over a given period of time' (Van den Top, 1998: 4-5). In this definition, the historical process of cocoa plantation in southeastern Ghana is a form of forest degradation as well. This definition is usually embraced by government forestry officers. The district Director of SODEFOR[23] is no exception, as he believes that 'agricultural activities are actually the main reasons of forest degradation in the district'. To him, farmers penetrate the reserved forests for farming, regardless of the interdiction from the forestry agency. From his perspective, the magnitude of the population growth exacerbates pressure on lands, for farming activities. This is why, according to him, forests are degraded by agricultural activities. Therefore, farmers are the ones who destroy forests, that is, the very basis of their livelihood systems (see Chapter 2).

During the interview, I realized that the director had not at all mentioned the rampant logging activities going on in the area. Yet, it is easy to indirectly observe the presence of timber companies in the region, witnessing the many big trucks transporting logs who drive past every time. Besides, the issue has also been raised in the FGDs and during interviews with local people. So, when I raised the issue with the district's forestry agency's director, he said: 'we do respect international regulations for timber extraction. The major reason of forest degradation in the Moyen Comoé region is agricultural activities, which are occupying 25% of the reserved forests, and not the logging activities'. Obviously, the local people have a different opinion.

People's own perception of the decline of forest resources, and of local ecological knowledge

> *'Forest degradation happens when you burn a forest, without planting. If there is no reforestation, then there is forest degradation. In my forest, I have some trees that are 200 years old. When you cut a tree, within two years, it grows up absolutely. Our ancestors burnt the forest without getting any forest degradation problems. We cannot destroy the forest because it is our treasure, the source of our livelihoods. If we destroy it, where will we go? In our view, the population growth affects the forests' sustainability in this area, together with multicultural interactions. In fact, the ecological knowledge of savanna people differs from that of forest people. Today, apart from the decrease of the rainfall we are experiencing, the migration of people with different cultures leads also to an interaction of ecological knowledges, which are not always the same. In this context, environmental management and protection become difficult. In practice, what happens is*

[23] SODEFOR is the national Forestry Agency, represented in different regions of the country.

> *this: we Agni, can cultivate banana, coffee, and cocoa in the same space with trees. However, to cultivate rice, which is the staple food of most of the migrant people, you must cut all the trees for the rice field. So, one of the problems we have to protect our environment here is, to some extent, due to the problem of the management of different cultivation systems and different ecological knowledge in the area. However, for us, logging activities remain the main reason of forest degradation here. These activities are performed by people with the consent of the state itself. The "forest exploiters" enter the reserved forests with the license from the state to cut trees in the forests at their convenience. So because of these logging activities, we have less and less virgin forest in the area' [Kakou, Pagnan, 15-02-08].*

From the information obtained through the FGDs and interviews with local leaders, it became clear that they have seen signs of land degradation and forest degradation during the past six years. According to them, since six years, the water from the well has not been as abundant as it was in the past. During the dry season, i.e. from December to March, it is muddy. Compared to the past, there is now lack of fuel wood and snails in the fields. At present, women have to go very far to fetch fuel wood for cooking.

> *'In the past, there were plenty of snails in the fields. We were eating snails to compensate for the lack of protein in our food. Today, we can catch less snails in the fields, compared to the past. It is not raining like it did in the past. Compared to the past, the climate has changed so much. We can see the climate change impact on our agricultural productivity. People in this area are always complaining about the degradation of lands and a severe decrease in food production and income as well. However, we are strongly opposed to the forestry agency's statement on land degradation and forest degradation. In their view, we are the key actors of forest degradation, because of bush fires we provoke when doing culture sur brûlis, slash and burn. But we cannot stop the culture sur brulis, because for us, it is the best and cheapest way to clear(ing) the field for agricultural production. Otherwise, cleaning the field without burning is very costly for us, since we have to hire labour for that. We don't have such money.' [Kakou, Pagnan, 15-02-08].*

Farmers, when they speak of forest degradation based on their local ecological knowledge, contradict the definition and accusations of the forestry and extension agents, which state that swidden cultivation leads to and exacerbates soil erosion and negatively influences food productivity. They believe that without the practice of swidden agriculture, insects would penetrate the yam and banana and relentlessly spoil their food crop production.

The local community organization leader for forest protection stated:

> *'One day, with the ANADER, the extension agents, we agreed to compare the efficiency of both systems of cultivation, that is, their system of cultivation with and without burning. Honesty, we won. The field cultivated with our local system produced nice and big bananas and yams, whereas their field failed to produce very well, as we had expected. In spite of that, they are still focusing on their technique; we are not ready to give up our technique either, since we don't have the money to respect their requirement for our agricultural production. From our agricultural knowledge, we know that bananas and yams need a land containing potassium, and you can get such land only with slash and burn. Moreover, we have our own system to assess the soil fertility with regard to the cash or food crops we want to grow. For instance, when the soil is soft, it is good for cocoa production. When it is too soft, it is not good for yam production. When the field is burnt too severely, you cannot get a good production of bêtê-bêtê, but must grow the n'za (the soft one) variety[24]. So, from our perspective, they will have to check the reasons for forest degradation, beyond the system of cultivation with slash and burn' [Kakou, Pagnan, 15-02-8].*

However, local people acknowledge that when the fire they provoke for burning fields is not controlled, this can lead to extensive bush fires. In the past, their parents did not have the problem of bush fires and their control because it was raining regularly. Nowadays, they acknowledge that it has become a problem. According to the local leader for forest protection, in 1960, sixteen million acres of forest disappeared in Côte d'Ivoire due to bush fires. In 1983, the eastern region knew strong bush fires, which has led to the loss of 4,000 acres of cash crops and food crops, for 390 farmers. This impacted strongly on the prosperity of the region, which was known in the past as the leading producer of cocoa and coffee in Côte d'Ivoire. This natural disaster motivated the creation of a national NGO against bush fires, called *Ebo Agnan Iti*, which means: 'the forest is secured'. In 1998, the NGO has implemented a sensitization programme in the villages surrounding Y.F. and beyond the region. Forests protection actions are performed during the dry season, from December to March. The NGO set up the *Comités Villageois de Surveillance* (village committees for forest protection), composed of young people whose role is to secure the forests. Local stakeholders involved in the NGO came up with rules to protect the forests, with the support and blessings of the traditional chiefs in the region. According to these rules, nobody is allowed to take along a match when he is going to farm or to cook in the forest. After 10 o'clock in the morning, the forest guards have to patrol to see whether someone is cooking. If someone is arrested with a match or is cooking in the forest, he is fined and has to pay 50,000 FCFA. If someone provokes bush fires, he has to pay 20,000 FCFA. If the bush fires reach the field of someone

[24] *N'za* and *bêtê-bêtê* are both varieties of Dioscorea rotunda. *Bêtê-bêtê* in Ashanti language means the soft one (Visser, 1975: 28-29).

else, the culprit has to pay 150,000 FCFA, which is the price for a cow, plus a box of wine, two sheep, and two chickens. Actually, they stated it has worked out very well because the district chiefs and the village chiefs were strongly involved in the agreements, and a sacred pact has been signed between villages to implement and respect the rules and sanctions.

> *'Today, the outcomes exceed our expectations. The reforestation in the area is now visible. The green vegetation is reappearing. So, for us, the outcomes are positive. We can state that we are achieving our goal, "Ebo Agnan Iti": the forest is going to be safeguarded' [Kakou, Pagnan, 15-02-8].*

The NGO now wishes to extend its experiences on forests protection and security to the other regions of Côte d'Ivoire, but is still lacking the financial means to achieve this ambition. However, it has organized sensitization activities in 12 districts: Agnibilekro, Bondoukou, Daoukro, Bocanda, Dimbokro, Toumodi, Bongouanou, Agboville, Adzope, Tiassale, Divo, and Oumé. Furthermore, SODEFOR has sent some farmers to the NGO to be trained and get support, to set up a similar NGO in their area, called *Bomangaouni*.

In sum, local people acknowledge that they possess a local ecological knowledge to protect their environment. What's more, from an Agni cultural perspective, forest degradation does not exist, to some extent, because for them, 'when a field is burned, it is to plant'; they do not burn without planting.

The exchange of ecological knowledge between forestry officers and local farmers has been widely discussed by Fairhead and Leach (1998). Discussing the anthropogenic influences on vegetation, Fairhead and Leach (1998: 186-187) believe that 'the enduring effect of settlement on vegetation is fundamental. Inhabitants in their everyday lives can render soils highly fertile, and manage village peripheries in ways which shape tree cover. Historically fortification has been important, but land management for kitchen gardening, multi-storey home gardening, fire protection and ritual purposes have also served to create patches of forest in the savanna ... The effects endure long after a settlement site is abandoned, while evidence of such processes can be found in savannas and forests throughout the region, it has consistently been ignored in analyses which have used these rich and impressive vegetation patches to represent the region's natural vegetation. Farmers frequently encourage particular tree species to grow in their fields or settlements, preserving, planting, or transplanting them whether for their economic values, beneficial effects on soils and crops or ritual purposes. Vegetation composing is frequently the outcome of such active management of particular trees'. In short, because farmers have to rely first and foremost on the environment for their livelihood, as well as on cultural rituals, they cannot be the destroyers of what they have to rely on for living. For generations, farmers have applied the local ecological knowledge and practices to deal with forest fires and soil erosion, and they have shown the capability to sustain the environment.

6.2.2 Migration, forest degradation and local ecological knowledge

During interviews, Agni men and the leader of the community organization for forest protection brought forward the interfaces between Agni and migrant ecological knowledges as an additional issue contributing to the forest degradation in the area. For the autochthonous Agni it is clear that migration and a culturally different interaction with the environment has an impact on forest management. Since migrants are mostly originating from savanna regions, they are obviously more knowledgeable about savanna management than about forest management. For this reason, they may destroy the forests while cultivating their basic food crops, which are dry land food crops like cassava and maize, in a forested environment. For instance, the Agni do not need to cut natural trees in their cocoa and coffee plot, because these plantations have already been developed by earlier generations, who have selectively cleared the forest (Van den Breemer, 1984; Hill, 1963) for cultivating banana, cassava and other crops. Migrants, on the other hand, do need to cut forest trees if they want to plant maize and rice, which are their staple foods.

With regard to solutions to overcome forest degradation, it is important to note that cultural practices indeed exist to protect forests. The *forêts sacrés* (sacred forests) are known to include protected species and spaces or areas that people are not allowed to make use of at all.

> *'Sensitive persons can see strange things there, when they enter these forests. In some of them, you can see a red cock, meaning the presence of ghosts or spirits. We have some rivers that never dry up, because they belong to the spirits. We don't touch them' [Kakou, Pagnan, 15-02-08].*

Moreover, according to the Agni, a *fetish* or spirit can protect forests that are not sacred forests. Notably, when people see a fetish in a forest, they fear to come near. Thus, the fear of fetishes makes people reluctant to go into primary forest. Furthermore, Agni men acknowledge that it is necessary to make the migrants aware of the need to stop cutting trees for the production of vegetables, because tomato, for instance, can be cultivated in secondary forest and previously used fields. There is no need to cut trees to open up a new space for such cultivation. For the Agni, reforestation is important to counter forest degradation. They have recommended that people get education about planting plants such as *Glycidia*, which grows rapidly and very much helps to fight soil degradation.

One of the traditional ways to address drought due to ecological instability is the traditional Agni witch dance, called *Moumoumin*, which is performed by women.

> *'Once this dance is performed, the rain will come. However, the practice is becoming more and more neglected, due to the impact of religion on people's lives. Our cultural practices are mostly forbidden by the religions. So, we can state that one of the reasons of our biodiversity's degradation is the lack of respect for our cultural environmental beliefs, values and practices. For instance, according to our culture, kings and chiefs are not allowed at all to fetch snails in the field. But today, some chiefs don't respect this cultural rule. So, we are now complaining about the scarcity of snails in the fields' [Kakou, Pagnan, 15-02-08].*

Although for the Agni population, growth and people's migration lead to different cultivation systems and ecological knowledges that have an impact on forest sustainability, they believe that abusing the forest by logging activities remains the major cause of forest degradation in this area. They are pessimistic about the impacts of reforestation activities on forest sustainability, because the *exploitants forestiers* are still continuing their activities. For this reason, I concur with Fairhead and Leach (1998: 193) that one should not deny 'the need to address major forest degradation problems in particular places, whether linked to timber extraction, mining, commercial plantations or farming. Rather, it does suggest the urgent need for some changes of approach, some more practical, and others more conceptual and indirect'.

6.3 Health problems and AIDS

From the five FGDs, interviews with the health center practitioners, Agni and migrant local leaders, key informants, as well as participant observation, it has been apparent that people have very little access to medical services. The lack of a medical infrastructure and facilities particularly affects the poorer segments of society, among which are women and children, and considerably increases their vulnerability to ill health. The lack of government support to make health facilities accessible to people who are already vulnerable, is part of what we consider to be external vulnerability. In turn, this external vulnerability jeopardizes the agricultural labour force for food crop production as well as food and livelihood security.

6.3.1 The health situation and women's reproductive health

According to the head doctor of the local health center, besides AIDS, the major diseases that people are faced with in the area are malaria, typhoid, abdominal pain, appendicitis, hernia, and children's anemia due to malnutrition. The services of the health center located in the village also extend to three neighbouring villages, which are situated two kilometers to the north to 26 kilometers to the south and six kilometers to the west.

According to the doctor, women and children are increasingly suffering from illness. Women have serious reproductive health problems. Many of them do not practice family planning, especially with regard to spacing. The gap between children's births is very small. Actually, this is one of the reasons of children's vulnerability in this area, affirmed the doctor:

> *'The reproductive health problem is one of the biggest problems in the area. There is no follow-up of the pregnancies. We advise four prenatal consultations. Women do on average two consultations. Even more, they do not use the medicine prescribed. For instance, among the medicines, there is an anti-anemic medicine costing monthly 1,200 FCFA, so 10,800 FCFA for the nine months. Unfortunately, they cannot afford even this medicine. Furthermore, not only do they have to pay for the delivery kit, which costs 15,000 FCFA, but they also have to pay for some medicines during the post-natal treatment of themselves and their babies' [Doctor, Y.F., 15-05-07].*

The FGDs with Agni and migrant women and the in-depth interviews with local leaders show that pregnant women cannot carry out the appropriate medical protocols because they lack the money to do so. Therefore, during their pregnancies and for the delivery they mostly rely on traditional medicines and healers. According to the midwife of the health center, many women in the area prefer to deliver at home rather than in the hospital. She believes that not only the lack of money has a negative impact on women's reproductive health, but also the lack of infrastructure (poor transportation). During their pregnancies, women from remote settlements have to walk, because they cannot pay the transportation cost to get to the hospital. As a result, they often reach the hospital in a very bad condition, with a fever, and anemic as well. Talking about these women's main health problems, the midwife states that they have lumbar pains because of the long distance they have traveled to get to the hospital. These women's lumbar pains have a strong impact on their pregnancies, she said. It can lead to a spontaneous abortion during the first trimester and to a premature delivery during the second trimester. Because of these problems, many women do not show up for prenatal care. 'They come only to get the medical book (*carnet medical*)', said the midwife. There is no follow-up of their pregnancies as the women do not respect the medical protocols, i.e. the medical visits and the medicines prescribed. Once they have delivered at home, women do not visit the hospital for the postnatal care, since to do so, they have to walk again with their babies, as far as six to 26 kilometers for those who live in other villages. Furthermore, when they can pay for transportation, they still have to walk some distance, since the taxi station is about 600 meters from the hospital. Consequently, in this area, most of the time babies remain unvaccinated, because women cannot meet the appointment scheduled each Wednesday for babies' vaccination. Women and their babies strongly suffer from the lack of prenatal and postnatal care. Pregnant women get serious blood pressure problems, eclampsia, premature abortions, anemia, and complications during delivery, which may cause stillborn babies.

> *'We are experiencing many cases of death here, because of transport problems. What's more, we don't have an ambulance to transfer the sick to Abengourou, the central town, particularly for those in need of surgery' [Midwife, Y.F., 15-05-07].*

At the state level, the alternative would be to implement a training course for traditional midwives in some remote areas, to assist women during delivery. However, these traditional midwives do not have any equipment, or an appropriate space for delivery. They deliver at home, often without light or electricity at night. They use fuel lanterns. During in-depth interviews, one of them acknowledged:

> *'We have no material to work with during a delivery. We don't even have a table. We are volunteers, without any stipend to motivate us. Coming from remote settlements, some women have to walk several kilometers to come to us to give birth. Often, they arrive bleeding. There is no ambulance, nor a taxi to reach the hospital. Therefore, most often, they are sent to hospital by motor bike or by bike. This exposes them to a uterine break, which requires surgery. Here, many women die from problems linked to pregnancies' [Ber, Y.F., 15-05-07].*

During FGDs, both Agni and migrant women complained about the cost of medical services, combined with the costs of transport for women living in remote settlements.

> *'The hospital is too expensive for us. How can a pregnant woman coming from a remote settlement afford a scan costing 10,000 FCFA and pay for the transport to go to Abengourou [the central town], where it can be done [the health center doesn't have an x-ray service]? So women prefer not to go, as they do not have the money for prenatal and postnatal care. Therefore, here young girls mostly go for an abortion with possible complications, without any care' [Akouba, Y.F., 27-04-07].*

In addition to what I have learnt from the health professionals at the hospital about the health problems in the area, I inquired about the meaning and interpretation of diseases and AIDS in Agni culture, and also about the views of the traditional healer on the health problems in the village.

6.3.2 Illness and AIDS in Agni culture: the *komian*'s view and practices

Some studies point out the role and importance of the *komian*, the traditional healer, in the Akan organizational structure (Rattray, 1959; Amon, 1960; Duchesne, 1994).

The *komian* is the healer, the protector and the advisor in Akan society. He/she is the priest who is knowledgeable about the 'fetish', that is, *suman* (in Ashanti language) and *amwan* (in Agni language), from which he/she gets the mystic power in society to heal and protect from devil spirits (Rattray, 1959; Amon, 1960; Duchesne, 1994). Rattray (1959: 23) defines *suman* (fetish) 'an object which is the potential dwelling-place of a spirit or spirits of an inferior status, generally belonging to the vegetable kingdom; this object is also closely associated with the control of the powers of evil or black magic, for personal ends but not necessarily to assist the owner to work evil, since it is used as much for defensive or for offensive purposes'. For Rattray (1959), *Suman*[25] relates to *nyame* (God), *abosom* (divinities) and *samanfo* (spirits of deceased people). It portrays an alliance between human and invisibles forces. There is not one single *suman*, but several. Each *suman* has a particular function and rules that allow the *komian* to perform his role in Akan society by interacting with *nyame, abosom* and *samanfo*. The *gyabom*, for example, which is a well-known *suman*, 'is a powerful charm for driving away evilly disposed and disembodied human spirits' (Rattray, 1959: 22).

In the research area, the *komian* informed me about the local meanings of illness, and the practices related to illness in Agni culture. Illness is called *awongnalé*. It is interpreted and supposed to be cured by the *komian*, the traditional healer who is knowledgeable about all kinds of health problems in the community. In Agni culture, there are two classifications of illness: illness can be *wafoutché*, meaning a short illness, such as a headache, or *wayoawougnalé* or *ehnwouhyaya*, which means long and persistent or chronic illness. According to this classification, HIV is classified in the second category and is called *babasso*, known to be the chronic stage of *dedié*, which are sexually transmitted diseases (STDs).

In addition to what the doctor stated when speaking about the major health problems in this setting, the following statement of the *komian* confirms women's vulnerability to AIDS, and children's malnutrition due to food insecurity:

> *'Actually, in this village, talking about illnesses, people suffer more from malaria, typhoid, itere (chronic malaria), backache, and sexually transmitted diseases. Regarding the typology of illnesses and sick persons who visit me for treatment, women are the most infected by 'chaude pisse', sexually transmitted diseases, which eventually turn into babasso, leading to AIDS. Children are mostly suffering from convulsion, caused by anemia. In my opinion, this is due to the fact that they are not eating very well and are not*

[25] The sumans are *sora boka* i.e. a masque or statuette: a design of a human made in wood. Most of the time, they are outfitted with plants, animal parts, string, make-up, etc. They are well placed somewhere in the compound or in the forest, to be worshipped, following rituals of animal sacrifices and the provision of food to the suman.

regularly dewormed. They are always bloated when they come to me. I use medicinal plants to deliver them from the worms in their bellies. I usually take four days for this kind of treatment. The child has to drink a medicine; I also scrape his body with another one, mixed with shea butter. However, I usually ask the children's parents to visit the hospital as well, because the medicine from there is effective, when it comes to treating anemia. We komian are knowledgeable about the causes of illnesses, and the adequate treatment for healing them. Through dreams, or when we are in trance, we get information from the spirits about traditional medicinal knowledge and guidance for healing. More often than not, in the case of ehnwouhyaya, the persistent illness, the spirits ask for the treatment first of all that a chicken is killed. Depending on the position of the chicken while dying, we can see whether or not the sick person will recover from the illness, whether the spirits accept to heal him. When the spirits agree, they show us the medicinal plants to be used and the offerings to be made. It is an ongoing process until recovery. Therefore, we are not the ones who are healing the people, but the spirits are, because they are the ones who show us the causes of the illnesses and the way to treat and cure them. However, speaking about AIDS, I did not yet meet a komian who successfully cured an HIV-positive person. We can actually cure someone from babasso, but not from AIDS. But we are committed to find a medicine against AIDS. We are working on that. In this area, we are with about sixteen komians. Most of them are women[26]*; maybe because women are closer to sick people, they also give birth and they are more concerned about protecting human life as well. Talking about myself, I was born komian, because my mother, who had difficulties getting a baby, worshipped a river spirit to be able to conceive me. Following the pact she made with the spirit, I had to become a komian. So even before I was born, my people knew that I would be a komian. When I won't be alive anymore, the spirits will choose someone among my children to be komian, to continue the legacy of my healing activities. We komians are always dressed in white clothes, because every time we see the spirits, they are dressed in white clothes. We get this recommendation from them to be always dressed in white' [Komian, Y.F., 16-05-07].*

6.3.3 Cultural narratives regarding AIDS and witchcraft

From an anthropological perspective, there definitely is a cultural meaning and interpretation of illnesses such as AIDS: 'a cultural model of sickness shapes the form and contents of people's illness narrative' (Radstake, 2000: 76). A prolonged illness, such as AIDS, is often interpreted as an unbalanced relationship between the spirits

[26] Duchesne (1994: 131) also found that a feminization of *komian* is taking place, i.e. many women perform among the *komian*.

and humans and among humans, which has come about through witchcraft. It is noteworthy that in Akan culture, a prolonged illness is sometimes interpreted as an outcome of the spirits' anger or the result of witchcraft. This means that a person can send you an illness or disease. In this manner, AIDS may be interpreted as a spiritual illness due to a curse from the spirits, or due to witchcraft In fact, 'witchcraft is central in Akan social life' (Radstake, 2000: 84). A witch is known to be someone from your family, very close to you, but often – in patrilineal societies – from 'the dark side of kinship', which would then mean from the female side (Geschiere, in Meyer, 1995: 271). 'Female relatives are more often suspected of witchcraft than men' (Bleek, 1975: 376). A witch is sent by someone's enemy, and witchcraft resides in jealousy and envy, which in the end may be so forceful as to cause the death of someone, by causing a disease. In a different cultural context, some people believe that if a witch does not like to see his or her enemy any longer, he/she will kill that person instantly and will not use a prolonged illness like AIDS to achieve his/her goal (Nombo, 2007). In my research area, some people believe that witchcraft is more often at the root of a prolonged and unknown illness. Still, in public, one does not speak lightly of witchcraft. Survey data show that 35% of the households that reported on the occurrence of unknown diseases did not mention witchcraft as their cause. Yet, when the question was raised during the FGDs whether there were linkages between AIDS and witchcraft, some people actually did believe that there is a connection between prolonged illnesses, among which AIDS, and witchcraft. They indeed assumed that unknown illnesses are likely to be witchcraft-induced.

In the research area, some people assume that the impacts on people's lives of modernity and the main religions have reduced the status and respect for the *komian*, who has cultural knowledge about unknown or witchcraft-induced illnesses.

Evidently, different causal narratives about AIDS can be valid at the same time for the same person. A person may accept the hospital diagnosis and at the same time believe his illness to be caused by witchcraft (Nombo, 2007; Niehof and Price, 2008). Still, 'a significant proportion of all AIDS hospital patients believe that counseling is irrelevant because the disease has been caused by evil spirits' (Caldwell *et al.*, 1993: 11).

In Côte d'Ivoire, like in many African countries, especially older women live in real fear of getting accused of being witches. In a context of witchcraft suspicion, women as caregivers may be considered dangerous to a sick person. Moreover, when women themselves are affected, they may not get any care from their family, as they might be treated as witches. The major reasons for women to get accused of witchcraft can be related to gender ideology and gender inequality. As they are considered to be more susceptible, women are sometimes seen as an easy target for an evil spirit that may harm people and expose them to prolonged and unknown illnesses, like AIDS. During my fieldwork, I came across several older women who suffered greatly from a lack of elderly care, and I suspect that some of them were being accused of witchcraft, which would account for the poor conditions they were living in.

In line with the above, witchcraft beliefs might help to spread the AIDS epidemic, as some people prefer to go to the *komian* for her services, rather than going to the hospital for HIV testing and treatment.

6.3.4 AIDS awareness and the mitigation programmes

Our results demonstrate that presently is AIDS awareness in the village. This has come about mainly by the implementation of AIDS programmes, together with a strong awareness campaign. Today, some people do not mind going to the public testing and counseling, and use condoms. However, girls are still exposed to unprotected and unsafe sexual intercourse, and therefore potentially to AIDS. Data collected from the health center that were confirmed by my interviews with the village midwife, stressed the many cases of abortion among young girls from secondary school, which made for a sharp contrast with the amount of AIDS awareness that had been claimed by the local leaders.

Talking about AIDS mitigation programmes in the village, the doctor of the health center stated in an interview:

> *'Since 2 February 2007, we do a HIV antibody test and treatment in this hospital. We work with EGPAF (an American AIDS programme) and the programme of PEPFAR, in collaboration with ANADER. However, we are able to attract only a very limited number of people to the hospital. Through our Prevention of Mother-to-Child Transmission Programme, we make women aware of AIDS during prenatal consultations. Our aim is to motivate them for the testing. So far, we did test 409 people, and 22 are HIV-positive, 18 of them being women. In fact, the 4 men are the spouses who accepted to do the testing, after being informed about their wives' infection. We have many difficulties to get the spouses of HIV-positive women to come for testing. Most of the time, infected women do not want their partners to be informed about their infection, because of the fear of being rejected and stigmatized. This is one of the major challenges we still have to overcome in response to AIDS in rural areas' [Principal doctor, Y.F., 15-05-07].*

An in-depth interview with the leader of the local Committee of the Response to AIDS made clear that the stigmatization problem indeed remains the biggest challenge the programme has to overcome:

> *'I do psychosocial care for HIV-positive persons and for children made vulnerable by AIDS. In one of the households, the wife is HIV-positive. With*

> *the agreement of the husband, we did his testing and found him to be HIV-negative. Actually, this couple is living very well. They have five children. They agreed to have safe sex. The wife is continuing her farming. For now, she is not experiencing a problem of stigmatization, because nobody in the village knows about her affection. But the problem can emerge in the future. Because when the programme will be distributing the kits to children made vulnerable by AIDS, is it going to be done in private or in public? With a public event of kits' distribution to children made vulnerable by AIDS, the village will know of the infected households' [Lass, Y.F., 16-05-07].*

Although the AIDS programmes in the village are perceived as being participatory in approach, as they involve local communities in response to AIDS, from a gender perspective, we can still see some disparities between men's and women's participation in response to the epidemic at the local level. In fact, there are more men in the local Committee of the Response to AIDS than there are women, since one of the criteria to be a member of this committee is to be literate. In this way, women are excluded from being part of this team, because of their illiteracy.

6.3.5 The prevention of mother-to-child transmission programme

The prevention of mother-to-child transmission programme is part of the AIDS mitigation programme, which is implemented by the local health center. However, this programme has many difficulties, due to the fact that women discontinue their participation. Women in the programme drop out, because they lack the money to pay the transportation fees necessary to meet the appointments at the hospital for their treatment. As said earlier, women from remote settlements have to walk to get to the hospital. Most often, they cannot manage to keep following the medical protocols. Besides, they cannot afford the baby formula milk when they have to stop breast-feeding their babies, as recommended by the programme. Additionally, they fear exclusion and stigmatization related to their HIV-positive status. They are afraid to stop breast-feeding their babies; to do so would mean that their husband and relatives-in-law will ask for a justification. That women have difficulties in taking part in the prevention of mother-to-child transmission programme has more to do with their social and material environment than with their willingness, as was confirmed by the midwife in charge of the programme:

> *'A HIV-positive mother has to stop breast-feeding her baby after it has become six months old, and give formula milk to the baby. But women do not have the money to follow this medical protocol. The programme is not providing baby formula milk or food. Therefore, mothers continue to breast-feed their babies after six months, thus exposing their babies to the virus. To illustrate this: five HIV-positive women are currently continuing to breast-feed their babies. Among them, one has a baby of ten months old. She said that she*

cannot stop breast-feeding her baby, because she does not have any other alternative. She said she is obliged to do that' [the midwife, Y.F., 15-05-07].

In conclusion, risky behaviour related to AIDS might be engendered by social problems and poverty, the lack of medical infrastructure and transportation means, as well as the budget constraints of the AIDS mitigation programmes.[27] These are all part of what is called the external vulnerability context, which increases people's vulnerability to AIDS. When the government cannot respond to people's needs in terms of medical services, transportation infrastructure and care, it jeopardizes effective responses to AIDS and its related impacts on food and livelihood security.

6.3.6 Women's status, children's health problems and malnutrition

Food insecurity experienced by households is found to be the main reason for malnutrition among children. Our results demonstrate that the hard living conditions experienced by women have something to do with their children's health problems and malnutrition. I could observe that many children show signs of malnutrition; most of them have protruding bellies and red hair. During FGDs with Agni women and migrant women, they acknowledged not only that they cannot afford the food needed to feed their children on a regular basis, but that they are also unable to regularly visit the hospital center with their children, because they lack the money to pay for transportation, the medical services and the medicine. This observation was made by the head doctor of the local health center. According to the doctor, among the children who visited the health center in 2005-2006, more than 48% of the 1-5 year-olds were diagnosed to be suffering from severe malnutrition, due to a worsened nutritional status; 13% of the children were suffering from chronic anemia, whereas 34% were infected by malaria.[28] According to the doctor, the children's health problems are mainly due to the malnutrition they experience because of their household's food insecurity. Moreover, on the basis of her overall knowledge of people's health problems, the *komian* acknowledges, too, that children present all kinds of signs that confirm their malnutrition, caused by the lack of access to sufficient and appropriate food experienced by their households.

It is believed that children's health problems are significantly connected to women's status. The relation between women's status and children's health has been well-documented (Caldwell, 1979; Caldwell *et al.*, 1990, 1993; Desai and Alva, 1998; Quisumbing *et al.*, 1998; Smith *et al.*, 2003; Niehof, 2010). The higher the women's socio-economic status is, the healthier their children are. It is noteworthy that women

[27] AIDS mitigation programmes do not respond to the needs assessment of people in terms of social security. Affected persons do not benefit from a full set of care services, i.e. medical care and financial support to improve their food and livelihood security.
[28] The village's health center data, 2005-2006.

with a low status are women who are economically vulnerable, with less access to and control over household assets and resources, and weak decision-making power. Hence, with regard to their low status, i.e. their economic vulnerability, they have less access to a good diet and health services. This situation influences their own nutritional and health status, as well as that of their children.

In developing countries like Côte d'Ivoire, women still have difficulties to observe the protocols of prenatal care, which requires proper nutrition during pregnancy. The local health center data indicate poorer diets and a reduced use of vitamins during pregnancy and/or maternity. The doctor of the health center indicated that their inadequate diet is continuously exposing women to diseases such as diabetes, arterial hypertension, eclampsia, malaria, diarrhoea, and streptococci infections that often lead to anaemia and tuberculosis.

In an AIDS context, it is noteworthy that the lack of an adequate diet increases the risk of a vertical transmission of the aids virus from mother to fetus. Women with a good economic status have a better nutritional and health status, and they have access to good quality care, both for themselves and for their children (Desai and Alva, 1998; Smith *et al.*, 2003). It is believed that children of educated mothers experience a lower mortality than children of uneducated mothers do. Studies demonstrate that there is a strong correlation between maternal education and child health, assuming that investments in women's education are important in reducing children's malnutrition (Schultz 1993; World Bank, 1993). We expect maternal education to improve children's nutritional status and health, because maternal education is linked to women's socio-economic situation, known to be a determinant of children's health (Desai and Alva, 1998).

During the five FGDs as well as by participant observation, I investigated the connections between women's status, their children's health, and the overall nutritional status in the village. Women remain the central actors of the food production, food provision and food security. They are still providing the bulk of the food production, spending most of their individual income on food and most of their labour time on activities to safeguard the security of consumption for their children. More than men's income, women's income is known to be spent on children's daily needs; because of their reproductive and productive gender role, they remain closer to the children (Quisumbing *et al.*, 1998; Smith *et al.*, 2003).

In conclusion, our results show that households, and particularly women, experience external vulnerability. This is visible in the decline in food crop production, and in the lack of access to extension services, formal credits, markets and medical facilities. Additionally, the lack of good soil that is a sign of the ecological vulnerability due to forest degradation is an important part of the proxies of the external vulnerability. Nevertheless, it is important to stress that Agni and migrant women are not passive victims of the AIDS epidemic. Instead, they actively develop coping strategies through social networks whenever they lack financial or economic capital. Below, I will present the role of social capital in coping with external vulnerability. I will show how people may use their social capital as their major coping strategy to deal with the general external vulnerability context.

6.4 Social capital and external vulnerability

Regarding the magnitude of the internal and external vulnerability that households are experiencing, they may rely on social capital at the community level to mitigate their economic hardship. It has also been documented (Nombo, 2007; Nombo and Niehof, 2008; Karuhanga, 2008), however, that in an AIDS-related socio-economic context, it becomes difficult to maintain local organizations like credit groups and religious groups because of illness, labour shortages, and lack of money to pay contributions. This affects the sustainability of the social capital at the community level. In such a context, one can see that the community can no longer support vulnerable people. We can therefore conclude that social capital at the individual and household level influences social capital at the community level.

Both Agni and migrant men and women acknowledge that compared to the past, social capital is not so strong anymore. They believe that it has been affected by the AIDS-related economic hardship. However, although social networks do not provide sufficient support to help affected households to sustain their livelihoods, they do somehow provide support to cope with hardship. Our results demonstrate that, within the community, people's willingness to assist each other during funerals still provides some support to bereaved families. Those who are involved in farmers' groups benefit from each other by sharing agricultural activities. Besides, collective action such as people's organization and commitment to fight against bushfires helps them to protect their plots, i.e. their livelihoods. And although the women's saving group does not provide enough cash, it helps them to cope with the hardship. Therefore, through collective action and social collaboration, people use social capital to decrease their vulnerability. In fact, people's commitment to local action, which shapes social capital, may have economic outcomes in the long term. For example, the group of people committed as volunteers to protect forests, secure their plots and livelihoods by taking preventive action. Therefore, involvement and participation in the forest protection group has positive outcomes, not only for the individual, but also for the community as a whole. This positive effect of sociability shows how networks can be important sources of power and influence for challenging vulnerability. Bourdieu's concept of social capital (1985) is instrumental here. He defines social capital as 'the aggregate of the actual or potential resources which are linked to possession of a durable network of more or less institutionalized relationships of mutual acquaintance or recognition' (Bourdieu, 1985: 248). It is not naturally given, but constructed with deliberate participation through people's investment, and their commitment to institutionalize group relations and claim mutual profits out of it. In the case of the forest protection network, the influence has been visible in the demand they received from outsiders, to organize a similar group in another village, to protect their forests. Hence, social capital portrays non-economic solutions to social problems (Portes, 1998). People realize that they cannot rely on state support, and that they have to rely on themselves to protect and sustain their livelihoods. The food exchanges that peoples are still continuing, regardless of their vulnerability, is also an expression of social capital, in the sense that it contributes to sustaining and reinforcing social relationships.

6.5 Conclusion and discussion

6.5.1 The gendered external vulnerability context

External vulnerability and the related effects are experienced by both Agni and migrant households and particularly by women, who are more affected by external vulnerability than men because of their gender role as food providers for their households. Institutional and structural poverty emerged as the main causes of women's poverty and vulnerability, visible from the fragile soil conditions and the lack of investments in conservation techniques, the reduced crop production, the poor access to extension services, credit, and medical services, and poor transportation facilities. Vulnerability to AIDS works at the interface with other adversities, such as ecological vulnerability, apparent in the scarcity of good soils for food production. Soil fertility management becomes a profound burden for farmers, especially women, in need of inputs in terms of labour and cash. This study shows the indirect effect of forest degradation on women's vulnerability to AIDS, as ecological vulnerability reduces the availability of fertile soil and food productivity. Women who have nearly no means of living (internal vulnerability) and are needy with regard to the infrastructure and social services, are in a situation in which their exposure to AIDS risks (external vulnerability) may increase, even to the extent that some of them might engage in transactional sex as a coping strategy. Getting access to land, extension services, and credit are therefore crucial means to achieve an optimal food productivity and household food security.

6.5.2 Women's lack of access to agricultural extension and credit

Despite the fundamental role of women in agriculture and rural development, the situation of rural women in Côte d'Ivoire is still not adequately addressed in policy and decision-making debates at the national level. One can see that there is a problem of governance when it comes to address the problems women have in getting access to extension services and credit. Agricultural and rural development approaches are still rather gender-insensitive. The lack of access to ANADER extension services obviously impacts food security, at the household level and beyond. Divorced and widowed Agni women can rely only on themselves for the production of food crops to feed their children. This contributes to the feminization of agriculture: the increasing share of women in the supply of agricultural labour. Microfinance projects and extension services could be major means to reduce women's socio-economic vulnerability and promote food security. This study shows that microfinance projects and extension services designed to target women farmers are largely missing in Côte d'Ivoire. Even so, it is noteworthy that the rationale behind targeting women in the microfinance programmes is the key role they play in poverty reduction. Since rural women play a vital role in rural livelihood systems, 'improving their access to financial services is assumed to enhance their productivity, improve their economic power, and consequently improve the well-being of their families' (Bekele, 2010: 2).

Microfinance institutions operating in the research area as well as in many other rural areas are not addressing women's financial constraints in a way that stimulates their participation. The deposit and the interest rate are simply too high to be afforded by women farmers. Hence, only a small number of rural women apply for credit. Furthermore, it seems evident that women farmers cannot meet the expenses for the extension services for crop production, because of the structural constraint of a lack of collateral and the financial constraints they are experiencing in their everyday life. In fact, with regard to women farmers' socio-economic vulnerability and their key role in food security, the agricultural national policy should make available subsidies to provide for the seeds and technical agricultural assistance they need, for sustaining their livelihood and food security. I concur with Hulme and Mosley (1996) that there is an urgent need to reconceptualize 'finance for the poor'. The target groups of 'finance for the poor' should be those who suffer from social inferiority, powerlessness and isolation. In the words of Hulme and Mosley (1996: 125): those who 'experience relatively intense levels of deprivation and discrimination: women (particularly female-headed households); the very poor (particularly the assetless, the landless and those dependent on agricultural labour); and the disabled'. When the 'finance for the poor' misses its target groups it cannot decrease their deprivation.

6.5.3 The feminization of illness

In the research area, a kind of feminization of illness can be observed. As was confirmed by both the medical doctor and the *komian*, that compared to men, women have more health problems. They have more malaria and *babasso*. The latter is a sexually transmitted disease. The women who are HIV-positive cannot meet the treatment schedule because of transportation problems, as explained above. Secondly, they cannot respond to the medical protocols to stop breastfeeding after ten months because they lack the money to buy the formula milk. Moreover, to stop breastfeeding is not well-regarded culturally, so stopping would create the risk of stigmatization and abandonment by the husband. Under these conditions, they not only increase their CD4[29] but also increase their babies' vulnerability to AIDS. Besides, women mostly deliver at home, which may cause AIDS affliction of the women who assist at the delivery. In fact, while performing the delivery, the traditional midwife is not protected, nor those who do the laundry afterwards.

Based on the general health picture presented by the hospital professionals and the *komian*, women are evidently the most vulnerable in terms of health problems. I therefore recommend that the Ivorian national health and AIDS mitigation policies should become more gender-sensitive. Although the ARV treatment is, in principle, free of charge for HIV-positive persons, I have witnessed that in the rural areas they still have to pay 1000 FCFA monthly, as a contribution for the treatment. Unfortunately, AIDS-affected households have problems to obtain sufficient income to be able to contribute to their own treatment. In conclusion, it is obvious that poor rural women cannot afford treatments like scanning at a cost of 10,000 FCFA, or a delivery kit at 15,000 FCFA, or 500 FCFA for contraceptives to engage in family planning. Hospital care is not attractive for women, or for girls, because it is too costly. While girls are exposed to unsafe sex and teenage abortion, they put their reproductive health at stake and increase their vulnerability to AIDS as well. The national government should be knowledgeable about these 'women's problems'. When underestimating these issues, the government cannot adequately reverse the AIDS epidemic in the rural areas. This will go at the expense of the individual's and rural community's health conditions, and it will jeopardize food production and agricultural labour in support of the nation's food security.

6.5.4 Biomedical discourse and the cultural politics of AIDS

The *komian* shows that there is an indigenous meaning to illness and healing. The *komian* is also knowledgeable about the overall and specific health problems in the area. She knows that women are the most vulnerable to AIDS in this area, because they are the most often infected by sexually transmitted diseases, leading most of the time to AIDS. However, unlike what some other traditional healers pretend, the *komians* acknowledge that, so far, no *komian* has been able to heal someone from AIDS. Yet, they are knowledgeable about AIDS with regard to recognizing its symptoms, and try to find a medicine to respond to AIDS. According to the principal *komian* of the area, the *komians* are working on that. She insists on the role and the importance of the *komian* to interpret and overcome all kinds of diseases in the local community, including AIDS, from a cultural perspective.

[29] CD4 are cells or helper T cells that are important in one's immune system as they help the body fight infection and disease. These cells get infected when HIV enters the body. For someone who has a healthy immune system, the CD4 cell count ranges from 500 to 1,800 per cubic millimeter of blood. AIDS is diagnosed when the CD4 cell count goes below 200. However, if your CD4 count is over 200, AIDS can be diagnosed if you have HIV and opportunistic diseases such as tuberculosis and pneumonia.

This raises the debate on the cultural politics of AIDS, and highlights the problem of knowledge interfaces shaped by the everyday interactions in the rural areas between modernity and African culture and between global and local knowledge. From the traditional healers' perspective, the modern world has had an impact on African beliefs and values, and modern development has led people to disregard social and cultural obligations, such as deference to the *komians* and their practical knowledge.

There actually is a need to consider local cultural interpretations and responses to AIDS, since the dominant medical knowledge on AIDS in global policies and interventions relies more on medical-technical rationality and neglects the endogenous factors and local rationalities. Such an approach disregards local cultural responses to AIDS. We believe that, in order for policies and interventions to be effective, it remains important to integrate medical approaches with an assessment of the local or cultural contours of the epidemic that are relevant for more effective mitigation strategies.

Schoepf (2001) places the epidemic in an international, political context and links the way AIDS is understood to the history of the relations between Africa and Western societies. She says: 'Societal responses to AIDS, including disease control policies, are propelled by cultural politics forged in the history of relations between Africa and the West' (Schoepf, 2001: 338).

In line with the above, Nauta (2008) argues that pharmaceutical industries are the powerful actors who have a monopoly on the interpretation and treatment of AIDS in Sub-Saharan African countries. Thus, the power relations between Western and African countries have an impact on the policies and interventions aimed to tackle AIDS in Africa:

> *'Although huge amounts of money are nowadays available for antiretroviral treatment in developing countries, the power of the pharmaceutical industry to determine the type and the price of treatment that becomes widely available must not be discounted' [Nauta, 2008: 9].*

While anthropologists emphasize the importance of traditional medicine and healing practices, Schoepf shows that they sometimes fail to address the limitations of such medicinal knowledge and practices in relation to AIDS mitigation. Schoepf states that, from a modern medical perspective, traditional healers not only cannot heal AIDS, but they can even be dangerous, by giving the false impression that they actually can. To her, people are naïve when they rely on traditional healers to recover from AIDS.

People in rural Côte d'Ivoire do indeed go to the *komian,* in part because they cannot afford modern medical treatment. This has been confirmed by the medical doctor, who believes that the option of going to the *komian* instead of the medical doctor is not only due to cultural beliefs, but also to a lack of cash. However, although the *komian* is less expensive at the beginning of the healing process, in the end, going to a *komian* can turn out to be more expensive than treatment by the medical doctor because of the duration of the healing process. This research has demonstrated that, in the AIDS context, the option of going to the *komian* instead of the medical doctor has something to do with the fear of stigmatization. With the *komian*, people feel they are protected from stigmatization, because AIDS is interpreted as *babasso*, chronic malaria, or a disease induced by witchcraft. These are somehow acknowledged as 'normal' afflictions, unrelated to AIDS, the result of mistrust and social problems in the community. Consequently, the afflicted person is not stigmatized as (s)he is in the case of AIDS, and may receive better psychological support from her/his relatives and the wider community. In other words, where Schoepf looks at medical effectiveness, the *komian* has an holistic approach looking at the mind and body of the afflicted individual and the social environment. If medically, the *komian* is not as effective as the medical doctor, (s)he can be very effective socially and psychologically.

Claude Lévi-Strauss (1963) has highlighted the role a tradition healer can play in a community, with the effectiveness of symbols, beliefs and trust. Taking the case of a sick woman who was helped through the difficult labour of giving birth with the guidance of a *shaman*, he stated:

> *'The cure would consist, therefore, in making explicit a situation originally existing on the emotional level and in rendering acceptable to the mind pains which the body refuses to tolerate. That the myth of shaman does not correspond to an objective reality does not matter. The sick woman believes in the myth and belongs to a society which believes in it. The tutelary spirits and malevolent spirits, the supernatural monsters and magical animals, are all part of a coherent system on which the native conception of the universe is founded. The sick woman accepts these mythical beings or, more accurately, she has never questioned their existence. What she does not accept are the incoherent and arbitrary pains, which are an alien element in her system but which the shaman, calling upon myth, will re-integrate within a whole where everything is meaningful. Once the sick woman understands, however, she does more than resign herself; she gets well. But no such thing happens to our sick when the causes of their diseases have been explained to them in terms of secretions, germs, or viruses' [Lévi-Strauss, 1963: 197].*

I conclude that the conflicting situation in which traditional medicine and modern medicine are competing with each other, increases rather than reduces people's vulnerability to AIDS, particularly in the case of women. The biomedical discourse on AIDS should not be the dominant one. In local communities, it should deal in a collaborative way with traditional healers' knowledge to address health problems and treatment efficiency. It is difficult for sub-Saharan African countries to make any real progress in their responses to AIDS, because policies and interventions do not adequately address the endogenous factors that include the social and cultural practices, discourses and responses to AIDS in rural society.

6.5.5 Stigmatization and its consequences

Because AIDS-affected individuals have to be secretive about their affliction, to escape social marginalization and exclusion, it is difficult for them to observe the counseling and treatment protocols. For women, the lack of money and the stigmatization are actually challenging the prevention of mother-to-child-transmission programme in the area. Therefore, the programme is far from achieving its goal, to protect women from AIDS along with their babies. It is noteworthy that women's socio-economic vulnerability is also due to gender inequalities, which prevents them from accessing economic resources. Furthermore, they remain the ones who are the most exposed to rejection and stigmatization, because of the existing gender ideology. Women are afraid of informing their husbands about their affliction, because they fear abandonment and divorce.

I conclude that the prevention of mother-to-child transmission programme is gender-insensitive. As long as women are extremely poor, they will be unable to respond to the programme. The programme might be strengthened by providing financial and food support to affected rural households and women. Further, babies from affected mothers should get formula milk and food from the programme, whereas mothers should get grants from the programme as financial support, to improve their livelihood activities. Besides, the local community's AIDS advisors should be involved in the programme as well, to perform a follow-up at home, as they do for the PEPFAR-ANADER programme.

I believe that if women and men can be educated to improve women's share in household decision making, this could help women to overcome the stigmatization problem related to AIDS. Yet, this again requires much commitment from the community leaders as well as outside institutions to reduce women's cultural vulnerability.

Chapter 7.
At the encounter of the social: Agni and migrants women's resilience in a context of AIDS

This chapter introduces a theoretical discussion on human agency and social practices in order to set a general analytical framework to understand the agency, resilience, and endurance of Agni and migrant women with which they face the hardship they are going through. At this point, we may formulate the main question to be answered in this chapter: how do we analyze the capability of Agni and migrant women in using their kinship systems, i.e. how do they configure rules and resources in the course of their battle against vulnerability? How are we to understand the social processes through which Agni and migrant women use their kinship and agency to improve their social position? How do Agni and migrant women actively and passively identify with social spaces that are continuously created and reshaped through their everyday existence, which is made up of internal and external vulnerability? By drawing on the contributions of Giddens (1984) and Bourdieu (1977), as well as on the Wageningen actor-oriented approach (Long, 1989; Arce, 2003), I investigate the significance of Agni and migrant women's actions in their immediate society and structures. This chapter uses a focus on human agency to elucidate the relations between individual women, their households and communities, the wider social environment, the AIDS epidemic and its related impacts on food security in Y.F., and what this analysis tells us about AIDS policies and interventions.

7.1 Agni and migrant women and policies, programmes, and values

> *'We have created a cooperative in 2005, to resolve our problems of poverty, knowing that union makes force. We believed that together, we can get more power to overcome the economic hardship we are faced with, in our everyday life world. Hence, the name of the cooperative is Yeyobie, which means "we'll try to see". We engage in agricultural activities, in growing cassava and horticulture. At the beginning, we even made placali (mashed cassava for fufu) for selling, but this has increased our problems, since we still didn't earn back the money we need for refunding the debts we have contracted to finance the activities of the cooperative. From the past and from this experience, we acknowledged that we, Akan people, are not very knowledgeable about trading or business. Therefore, we prefer to produce food crops, and find people who will buy from us, for selling. We started our activities with an individual contribution. Each of us contributed 10,000*

> *FCFA. But the first project failed. Meanwhile, we decided to restart the activities with a microcredit we asked from a microfinance institution. However, we were supposed to make a contribution of 100,000 CFA before getting the microcredit, unfortunately we could not get such amount cash. As an alternative, we started an affiliation with FENACOVICI (a federation/ network of food crop production cooperatives), thinking that this could help us to manage our business of food crop production and marketing. Actually the experience turned out worse than we had expected. None of the promises we got from this network has been fulfilled. We went into debt because we borrowed money from people to manage our business*[30]*. Today, we are reorganizing and sensitizing the members of the cooperative to remain motivated. But the biggest difficulty is the members' contribution, owing to the increasing poverty among women. Nevertheless, we are currently farming one and half (1.5) acres of cassava. We intend to sell the production to someone who will take it to Abidjan, and use the cash to grow new field crops and refund our debts as well. It is now the season for clearing the fields, but we didn't get cash to do that yet. In the future, we plan to extend our field crops to farming yam, groundnuts, banana, cassava and horticulture. The queen-mother and the chief themselves already gave us land for farming. So, we don't have any land problem for farming. In the long term, if we get strong funding and support, we are ensured that the impacts of our activities could help us to feed our children, to send them to school, and to protect their health. Therefore, we believe that through agriculture, we can contribute to overcome the hardship we are faced with, but also contribute to the village's socio-economic development, by donations from our cooperative to support, for instance, the health center projects and the schooling projects in the village' [Catherine, Y.F., 28-02-08].*

Vulnerability is shaped by socio-economic, political, cultural and environmental dynamics. From the DFID perspective, policies, institutions and processes (PIPs) are the most important issues in the livelihoods and vulnerability framework, as they influence the availability of resources, and the ways in which people can use such resources to shape their livelihoods. Therefore, the PIPs are also perceived as transforming structures and processes. Private and public organizations implementing policies and delivering services definitely affect people's livelihoods, while processes configure social relations and institutions (Ellis, 2000). In the framework, the

[30] FENACOVICI is a national network of food crops production cooperatives. The network has been created to support rural women's livelihood activities, in order to reduce their vulnerability. FENACOVICI has a marketing background. The network sells food produced by the members, and gives them back the cash. Agni women of the research area who adhered to the network had a bad experience, as they never got the cash of their food produces sold by FENACOVICI. Therefore, their cooperative bankrupted, and went into debt, as they borrowed cash from people to manage their business.

interaction between the vulnerability context and the policies, institutions and processes (PIP) show that people may affect the vulnerability context directly or indirectly by shifting the PIP environment, but the effect may be limited (Brons *et al.*, 2007, Karuhanga, 2008). In this study, women definitely lack the power to make the institutions change. Women do not as yet have enough power to change the institutional context, but they do show resilience.

The findings of this study show that women's status is much influenced by cultural, economic, ecological, and health vulnerability, and a lack of relevant policies that obviously increases their vulnerability. However, in this vulnerability context, women do resist and may triumph over their poverty with three kind of coping strategies. The creation of a cooperative aims at earning assets and resources. For some women, engaging in transactional sex with the agricultural workers they hire is an alternative to overcome the labour shortages. Others have created new recipes to overcome food shortages.

The initiative to create a cooperative and to finance its activities did not come from outsiders, but from women themselves, determine to tackle their socio-economic vulnerability. Despite the failures they experienced with the first cooperative, they refused to give up. The case of some women leaders actually shows that vulnerability can be used as a positive power as well, when women use their agency to overcome vulnerability. This contrasts with a sustainable livelihood approach and vulnerability framework that usually present vulnerability primarily as something negative that has to be overcome by coping strategies. Here, we state that vulnerability itself can be a part of such strategy.

Drawing on an actor-oriented approach, Arce (2003) shows how policy-oriented, sustainable livelihoods literature has failed to recognize and understand the value contestations that shape the ways in which people interact with their livelihoods and cope with vulnerability. While having poverty alleviation for a starting point, along with reducing vulnerability, the sustainable livelihood approach failed to grasp the actors' reality, as it '... excludes what is important to the poor, i.e. reserves to meet contingencies' (Arce, 2003: 203). The results of the present study show the role and significance of Agni and migrant women's strategies and the decisions they make to maintain or transform their existing livelihoods to meet contingencies.

During FGDs with Agni and migrant women, I was surprised to see a consensus on what women apprehend as the main alternatives for their socio-economic development. Agni women are convinced that a relevant project of food crop production can help them to get out of poverty and ensure their households' food security.

> *'We, Agni women, what we know in terms of livelihood activities, is food crop production. Actually, if we have access to agricultural means and inputs, it*

> *is impossible for us to be poor while we are engaged in food crop production activities [Catherine, Y.F., 15-02-08].*

For migrant women, getting access to grants or credits to manage business activities, such as selling food produces and other provisions, is the only way to get out of poverty and to be able to support their households' food security.

> *'For us, djoulaya, or managing our commercial business activities, is what we know, and is for us the best livelihood activity we can do. Here, migrant women who rely only on cropping are the ones who don't have any other alternative; otherwise, they would go for commercial business activities. For us, it is the best alternative to get out of poverty [Kadi, Y.F., 15-02-08].*

Regardless of the failure of the Agni women's first cooperative, they maintain the same livelihood activities for the second cooperative, i.e. food crop production, but decided to leave out the marketing side, as they acknowledge that they lack good marketing skills, compared to migrant women.

One can see that food crop production for Agni women and business activities for migrant women are part of their identities. Denying them these activities, replacing them or imposing other livelihood activities for potential empowerment projects may constitute a threat to their identity. The Agni women made this value claim when they started another cooperative, with the maintenance of the same livelihood activities, i.e. food crop production. Even more, Agni women used their ability to contest by resigning from the FENACOVICI network (see above), because they did not get the outcomes they expected when they joined the network.

In short, while analyzing people's vulnerability and their coping mechanisms, one should look at their social values and interests, and 'how these interests are brought together to construct local claims, a variety of social relations, actions and language strategies' (Arce, 2003: 210). One should look at the actors' understanding of their own reality, and opt for 'a more dynamic approach to the understanding of local livelihoods which stresses the interplay and mutual determination of the contestations of values and relationships and which recognizes the central place played by human action and identity rather than capital' (Arce, 2003: 204). In this manner, Agni and migrant women's coping mechanisms depict the real context in which they live, and offer insight in the ways both Agni and migrant women use their power and agency. I concur with Arce (2003) that understanding livelihood realities means understanding the complex varieties of social transactions that cannot be measured as asset and capital. It is imperative to transcend a functionalist understanding of social reality which overlooks value contestations, in order to see how women use their power and agency for their livelihood generation, and how they use it to challenge their vulnerability.

In his address to the 88th annual meeting of the American Anthropological Association, Eric Wolf has shown how people can get power when they are faced with vulnerability, and how they may use it to deal with it: 'The arrangements of a society become most visible when they are challenged by crisis. The role of power also becomes most evident in instances where major organizational transformations put signification under challenge' (Wolf, 1990: 593). Challenged by economic hardship, women's capabilities to resist to cultural, economic, ecological, and health shocks become more visible. From an African feminist theories perspective, this exercise of women's power, unknown by 'outsiders', challenges Western global discourses on gender and women's vulnerability in the African context. According to Amadiume (1987), while addressing gender issues, Western feminists mostly depict African women as generally oppressed and powerless. As such, they do not look into the socio-cultural systems which guarantee women's power. For her, unlike what anthropologists and Western feminists have said about African women's lack of power, they do have power, embedded in their positions and visible in the magnitude of their duties within their society.

The battle of both Agni and migrant women to get out of poverty is an inherent part of their motivation; it is like an internal, spiritual strength and power they show in their everyday lives to challenge their impoverishment and continuously perform their responsibility *vis-à-vis* their households. They are constantly aware that their households' members rely on them in terms of food and health care. This responsibility that they carry on their shoulders generates a self-confidence and belief in their own capabilities to fight for their households. Finally, they are the ones who remain strong, the ones who have power, which is the power of resistance to vulnerability. Women battle at their own expense. We make them into victims, but they do not accept to be victims.

In other words, the way women fight against being vulnerable contrasts with the vulnerability discourse which victimizes women. Instead, they show in their everyday lives that they are very rational in their struggle against impoverishment. Their battle is dynamic, full of creativity, strategies, risks, negotiations, and regrets, including attempts to strike a balance between health risks and the economic survival of their households. For some of them, engaging in transactional sex as compensation and extreme alternative to pay for hired labour is actually part of the strategy to feed their households. Since livelihood studies are known to be very goal-oriented and asset-based, they do not render explicit the processes that shape these women's struggles to improve their livelihoods and overcome the stress and shocks that are an inherent part of it.

7.2 Agni and migrant women's agency and given structures

By inquiring into the nature of Agni and migrant women's agency *vis-à-vis* their kinship systems, we may come to understand how their practices and actions, stemming from their everyday life, shape and construct local structure, rules and norms. A focus on women's agency may lead to a better understanding of the social in the context of AIDS mitigation policies in sub-Saharan African countries, and particularly in Côte d'Ivoire. It allows us to look into the ways in which women as social actors interact with their cultural institutions, AIDS mitigation policies and interventions, which is missing in the AIDS policies framework. Hence, we need to draw on social theory in order to deepen our understanding of the theoretical meaning of human agency in contrast to structural approaches, which are emphasized by AIDS policies along with development policies and projects (Long and Ploeg 1989, 1994; Otsuki, 2007).

Coming back to the human agency of Agni and migrant women, one can understand the social process through which these women use their kinship systems to improve their social position. Following Giddens and Bourdieu, we are trying to understand how women actively shape or structure matrilineal and patrilineal relationships, we can identify their positions and dispositions within the social field. With regard to their dispositions and positions, Agni women have duties, positions, that they are entitled to claim: they have, in an exclusive way, the decision-making power to nominate and select chiefs and kings to sustain their royalty, and the traditional power that is inherent in Agni culture; they perform the *moumoumin* witch dance that portrays Agni women's role of purification and protection, acknowledged by the entire Agni community; they recognize the role of the *komian*, who are mostly women, elucidating women's power in Agni community as well. The hierarchical power of the *reine-mère* (the queen-mother) has the potential to make things change in Agni community. She has an exceptional historical and cultural position, and yet the queen-mother, like ordinary Agni women cannot escape the fulfillment of her reproductive role, which is to sustain her culture through either the matrilineal kinship system. Agni women are obliged as much as entitled to claim their positions within their community, which are internalized as *habitus*. The Agni kinship structure is therefore shaping as much as it is shaped by Agni women's agency, i.e. their interactions to reproduce the Agni social structure. When we go back to the vulnerability context, it is apparent that Agni and migrant women use their agency within their communities to challenge the hardship, in order to provide food for their households. Understanding Agni and migrant women's battle against internal and external vulnerability is crucial to our understanding of how rules and resources emerge from women's struggles and negotiations in their everyday life. Agni structure is created, maintained and changed through action (Giddens, 1984). For example, while from an Agni cultural perspective Catherine was denied land ownership, as this right belongs to nephews, she could challenge that same cultural rule to be able to feed her household. Thus, she used her position and agency to improve her situation by obtaining access to the portions of land belonging to her grandfather, although she was not culturally entitled to inherit them. This means that Catherine as an actor is capable and knowledgeable of configuring the rules and resources to claim land property rights. Her case shows that kinship structure can indeed be changed from within by individual actors.

However, beyond this capability to claim their entitlements, women may use their agency in such a way that, in some circumstances, it may imply a potential risk to increase their exposure to the AIDS epidemic, thus increasing their households' food and livelihood insecurity. The following is a clear example. To cope with the increasing labour shortages they are faced with, some women can afford to hire labour only if they engage in transactional sex with the labourers, by way of payment. Exposing themselves to the risk of an AIDS infection in this way certainly poses a cultural problem for both Agni and migrant women, as they risk stigmatization within their ethnic group, but they take that risk rather than to give up their ethnic identity.

7.3 Social and cultural embeddedness of Agni and migrant women's agency

As discussed above, culture as a factor of vulnerability plays an important role in this study. Although they share hardship and adversity in their everyday life, Agni and migrant women are sensitive to shock and stress differently in several ways. Even if Agni women have access to land, they are faced with difficulties in taking advantage of it, due to the agricultural labour shortage and difficulties in hiring labour. The migrant women do not have access to land and are faced with difficulties getting access to the food market due to the cost of transportation, and making a profit out of their food produce reselling business. Besides these factors, which increase Agni and migrant women's vulnerabilities, cultural factors related to their matrilineal and patrilineal kinship display potentialities that increase their vulnerability to the AIDS epidemic. How are these factors socially and culturally embedded and how do women use their agency to deal with these potentialities?

From a gender ideology perspective and with regard to the gender division of labour, producing, cooking and processing food, caring for the sick, playing a major role during funerals by cooking and taking care of people, are among women's key roles that are socially and culturally embedded in society. Women's value is derived from their ability to somehow manage the hardship of fulfilling these social and cultural roles *vis-à-vis* their households, along with their respective communities. For them, fulfilling these cultural obligations is definitely valuable. In finding ways to manage the hardship, Agni and migrant women do show resilience, in which their agency plays an important role. Their agency is also shaped by their emotions and personalities. In the multi-layered process of their resilience, Agni and migrant women definitely show conscious and purposive behaviour when they consider the appropriateness and efficacy of their behaviour in a certain normative and institutional context. What is more, they may deliberately challenge cultural institutions while using their agency to cope with food shortages.

This study shows that to respond to food shortages, women may deliberately renounce their 'cultural food habit', with practices that they have developed as food coping strategies (see Chapter 5).

> *'We became Dioula women (migrant women), since we now eat their basic staple food, I mean kabato and kokohndehn. We even have a recipe of kabato made from dry banana, to deal with the problem of seasonal food shortages in the village. So, the problem of food we are facing in this village allowed us to get many new recipes to overcome our hunger. We can call them the recipes of the food crisis' [Madeleine, Y.F., 27-04-007].*

By adopting Dioula food habits, that is, by borrowing food recipes from Dioula women, Agni women negotiate cultural boundaries concerning food, while this is apparently not affecting their cultural Agni identity. Following Bourdieu, *habitus* is the historical legitimation of present practices. However, there are limited options for change. Agni and migrant women both challenge their food *habitus* through coping strategies they have developed and shaped by cultural interaction. The migrants do now eat *akpessi*, which is not their own staple food, but a recipe borrowed from the Agni. The obligation to change food habits along with food practices cannot be explained by a structural view of culture. Rather, people change and negotiate their cultural practices and attitudes, within limits.

Furthermore, the motivation to hide their household's food insecurity (see Chapter 5), portrays women's refusal to admit their 'incapability' to improve their household food security. Feeding the household is a socially and culturally embedded women's concern, more than it is a men's concern. For both Agni and migrant women, fulfilling this principal gender role is related to pride and womanhood in their respective communities, unless they accept to be stigmatized as useless women. The shame related to a household's food insecurity is experienced more by women than by men. Women are culturally valued as food producers, food processors, food providers and caregivers. They will not give up their gender identity or renounce the opportunity to assume their gender values. Even if this implies a lack of control over their own sexuality, as it does for migrant women.

7.3.1 A child's future reflects the submissive behaviour of the mother *vis-à-vis* the father: endurance or resilience?

As discussed in Chapter 5, migrant women are unable to refuse sex to their husbands, even in case of a suspected AIDS infection. In the process of fulfilling a cultural and gender duty *vis-à-vis* their husbands and the community, migrant women do show the ability to endure, rather than showing resilience. In being submissive, they do not reduce their vulnerability to AIDS, yet they do make themselves less vulnerable and more valuable to their community. On the other hand, if they are not submissive, they accept stigmatization and marginalization within the community. From a patrilineal migrant women's cultural perspective, a woman's ability to endure the patriarchal domination of her husband is telling of who she is, i.e. her position and her value in migrant society. A woman who endures, deserves respect and a blessing, not only for herself but also for her children. Hence the saying that 'a child's future reflects the submissive behaviour of the mother *vis-à-vis* the father'. Or, 'if your life is made of too many troubles, just ask your mother for the reasons'. Another saying is 'not every woman can bear a king'. Only women who have the capability to endure hardship in the household do. These are some sayings in the migrant community that portray the gender ideology and the rationale behind women's status and position in society. Therefore, one can see that, with regard to their own sexuality, whatever their agency, migrant women cannot do much but endure (cf. Matinga, 2008).

7.3.2 Culture, secrecy about AIDS, and resilience

The literature shows that the secrecy and stigma related to HIV and AIDS are part of the cultural and contextual vulnerability that definitely affect the agency of the affected people (Nombo, 2007; Nombo and Niehof, 2008). People refuse to acknowledge their HIV infection and prefer to seek the cause of their affliction in witchcraft. In doing so, they follow the common socio-cultural perception, interpretation and discourse surrounding AIDS, as a disease that lies more with malediction or punishment in case of someone's deviation from the cultural norms and values. This study shows that persistent illnesses, such as *babasso* and AIDS, may also be interpreted as a witchcraft-induced illness, which more often motivates people to call on support from the *komian* than on help from health professionals. In this way, a HIV-affected person receives emotional support from the community and, to some extent, gets the social space to enhance his/her resilience to deal with the disease and challenge its devastating impacts. However, I concur with Niehof (2008: 222) that 'measuring resilience is difficult, but in the case of resilience to HIV/AIDS even more so, because of the complex temporal and spatial dynamics of the epidemic'. With regard to the temporal perspective, this also applies to the present study of the Agni and migrant women, because the complex temporal dynamics of the Y.F. case have not been covered by a longitudinal study, in which data are collected over a longer period of time. Instead, I documented women's responses at the time of the research to the AIDS crisis that had developed over the previous years. Still, these responses, however restricted, do reveal resilience, endurance, and agency.

7.4 Agni and migrant women's agency and the vulnerability context

The previous Chapter 6 shows that external and contextual factors have effects on Agni and migrant women's agency, with regard to their efforts to fight vulnerability. The lack of infrastructure, like markets, health care facilities, extension services, access to formal credit, and environmental degradation, undoubtedly increase Agni and migrant women's hardship and restrict their ability to mitigate the effects.

Although women manage their reproductive health problems and keep bearing children with the support of the *komian*, relying most of the time on traditional medicine for pre-natal and post-natal care, it is obvious that the lack of access to reproductive health care affects their reproductive health along with the health and the future of their babies. HIV-affected women who are formally in the prevention of mother-to-child transmission programme protect themselves from stigmatization and potential divorce or exclusion by hiding their condition to their husbands and kin. Obviously, they may thus increase their babies' vulnerability to AIDS, and spread the disease among household members as well. The programme itself, however, has its limitations as well; it does not support affected mothers financially, which keeps them from providing formula milk to their babies, as required by the treatment, and which adds to the babies' vulnerability to AIDS.

The lack of access to extension services clearly affects food production and household food security, and may potentially increase women's exposure to transactional sex. Therefore, due to the lack of cash some women may engage in sex with the hired labourers by way of payment for their labour. Although they cannot get a consistent profit out of the marketing of food produce due to difficulties to get access to markets, migrant women will not give up. But they are aware that their resilience against AIDS remains seriously influenced by the institutional context. Moreover, while Agni and migrant women cannot get access to formal credit due to the lack of cash to pay for the required collateral, they use their creativity by engaging in informal micro-credit arrangements among themselves to counter their condition of vulnerability. Yet they acknowledge that with the lack of access to formal credit they cannot challenge their condition in a consistent and sustainable way. It has also become obvious that the environmental degradation, especially of the forest and soil quality, has the potential of increasing their vulnerability in the future.

7.5 Social capital and households' resilience to AIDS impacts

Studies show that social, context-related AIDS impacts are likely to negatively affect people's vulnerability (Donahue *et al.*, 2001; Niehof, 2008). In this study, during the FGDs with Agni and migrants, people have reported a decline of reciprocated help in the village and among kin. Both Agni and migrant households acknowledged difficulties in meeting the obligations that are required to sustain their membership of social networks. The cases of Cathi and Ama (Chapter 5) show the lack of time women have to attend meetings in the village as well as the lack of cash to pay their contribution, due to the erosion of household assets due to health-related adversity. Social capital may be fragile because of the lack of material means to nurture it, especially to reciprocate, in the sense that people may no longer support each other to get out of the hardship, since the crisis may affect them equally and simultaneously (Putnam, 2000; Nombo and Niehof, 2008). Even though a household may get support from kin, neighbours, or other networks to finance a medical treatment, the support may not be as reliable as it was in the past, as households all share the same fate of poverty and difficulties to invest in social relations. Therefore, I concur with Nombo and Niehof, (2008: 253) that 'moral principles of solidarity and mutual help – also between kin – break down in a situation of shared hardship, leading to depletion of social capital. Lack of resources to invest in social networks and groups make individuals unable to generate and use social capital that could help them become resilient to AIDS impacts and other shocks and stress'. However, this study also shows different results particularly in the case of people who purposely invest in keeping up the social capital through their continued participation in funerals, despite the fact that the rising number of funerals due to AIDS-related morbidity increases the burden.

Compared to migrants, the Agni are more often confronted with the cultural burden of funerals. Despite the poverty in the village, they have to reciprocate, by giving their financial contribution to any funeral of their Agni co-residents, both within and outside the village. Women experience the burden of the contribution to funerals more frequently, as they have to pay for themselves and for their children, often on top of their contribution to the cooking and taking care of the guests. In spite of their poverty, women manage to fulfill this cultural obligation. However, unlike what Nombo and Niehof (2008) found, this study shows that in a situation of widespread poverty, the social obligations embodied by social institutions can still be respected. The case of the Agni of Y.F. portrays this fact and shows the importance of funerals in Agni culture (Eschlimann, 1985), even in the context of AIDS.

7.6 Conclusion

Agni and migrant women emphasize their intentionality through their behaviour (Giddens, 1979) as a routine feature of their everyday life conduct, in the process of their struggle against vulnerability, and enduring hardship to retain their own social and moral values in their respective communities. Agni and migrant women engage in conscious and purposive behaviour. In other words, they are actors who believe in the appropriateness and effectiveness of their behaviour in the institutional context of Agni and migrant culture.

Paying attention to the agency of Agni and migrant women helps us to understand that the active structuration of social relationships involves the interdependency of Agni and migrant women's individual agency and the social-cultural institutional structures, like matrilineal and patrilineal kinship organisation. The rules and resources organized as properties by Agni and migrant social organisation are both the medium and the outcome of the practices in their respective communities that constitute those systems. So, Agni and migrant institutions are the outcome of Agni and migrant women's actions, since their social systems are historically reproduced by their actions. Although Agni matrilineal and migrant patrilineal structures to a large extent constrain Agni and migrant women's actions, these women also show the power of resistance and the power of endurance *vis-à-vis* adversities in their everyday life. The solutions they find to their problems are visible in their social interactions, like their food production and consumption strategies, the idea of running a crop production cooperative for Agni women, their reliance on the *komian* and traditional medicine, and the arrangement of informal micro credit among themselves. This chapter has shown that vulnerability does not necessarily portray a weakness, but that it can also be positively used as a part of women's strategy to counter adversities.

Chapter 8.
Answering the research questions and policy implications

This study focused on the gender aspects of the impacts of AIDS on food and livelihood security in rural Côte d'Ivoire. The study concentrated on three main aspects. First, using data from a village in southeastern Côte d'Ivoire, I investigated how cultural features such as matrilineal and patrilineal kinship systems contribute to the increase of women's exposure to the AIDS epidemic, and how this affects their food and livelihood security. Secondly, by drawing on household data, I investigated how the internal and external vulnerability lead to differentiation among autochthonous Agni and migrant women. Thirdly, I described the role of women's agency to cope with their affliction, as well as women's resilience in fighting vulnerability. Lastly, I sought to understand whether AIDS and food policies in Côte d'Ivoire are efficient enough to challenge the devastating impacts of AIDS on food and livelihoods security, especially in the rural areas. This final chapter presents the conclusions and discussion of the main findings with regard to the objectives of the study, followed by the policy implications and recommendations for future research.

8.1 Summary and discussion of the main findings

8.1.1 Culture matters in AIDS mainstream knowledge, policy and intervention

The first objective of this research was to provide insight into how cultural features such as matrilineal and patrilineal kinship may potentially contribute to increase women's exposure to the AIDS epidemic. Issues on the differential vulnerability with regard to the cultural systems of Agni and migrant women have been discussed in Chapter 5. It shows that matrilineal and patrilineal kinship provides different potentialities and constraints for Agni and migrant women, respectively. While potentially some Agni women may engage in relations with different partners to perform their reproductive roles, patrilineal migrant women have no control of their sexuality before marriage, while afterwards, they have to be submissive to their husband's sexual desires, even in the case of an AIDS infection, at the risk of becoming stigmatized in their community for not complying with cultural norm about gendered behaviour. Many of the studies that contributed to develop AIDS mitigation policies and programmes do not adequately address the socio-cultural and gendered issues underlying the scope of women's vulnerability to the AIDS epidemic as compared to their male counterparts, particularly in Sub-Saharan Africa.

Most of time, these studies do not explore women's vulnerability to the AIDS epidemic and the role of their agency in coping with the impacts that emerge as a result of their vulnerability to the disease from a cultural perspective. My research findings indicate that besides matrilineal and patrilineal kinship systems, other cultural features motivate the secrecy related to HIV infection as well. Examples are cultural practices like the levirate and sorarate in migrant communities, which play an important role in the spreading of the epidemic among household members. People aggravate their vulnerability to the disease, alongside with its impacts on their food and livelihood security, by their reluctance to acknowledge their HIV infection and their preference for the secrecy of treating it as an affliction through witchcraft instead of publicly undergoing biomedical treatment. They do so because of the negative socio-cultural interpretation and discourses surrounding AIDS in their society, in terms of a disease caused by malediction and punishment. Despite the fact that the *komian*, who is culturally appointed to treat all kinds of diseases, does not show the capability to heal people from AIDS, she remains a strong emotional support for HIV-affected people and likely strengthens their resilience to challenge the disease.

This research contributes to the debate on the cultural politics of AIDS and questions whether mainstream explanations of the epidemic and the existing biomedical medication policy are efficient enough to respond to it in African countries like Côte d'Ivoire, because they do not address cultural and endogenous dynamics that may explain the spreading of the AIDS epidemic, especially among women. In this regard, I concur with Nauta (2008) and Schoepf (2001) who argue that the fact that AIDS is still challenging the world-wide efforts in African countries has something to do with cultural politics or the power relations shaping mainstream knowledge and global discourse, which rely more on explanations in terms of structural economic factors than socio-cultural and endogenous factors. In other words, the cultural politics of AIDS do not give enough space to developing countries to decide on their responses to the epidemic. The research demonstrates that there is a need to go beyond the 'development industry', driven by the international political economy, which makes it difficult to mitigate the disease on the continent. Research findings show that the refusal to include, on a context-specific basis, the site-specific social and cultural factors that underpin women's vulnerability to AIDS in the generally communicated knowledge, explanations, policies and interventions regarding AIDS, will continue to make it difficult to reverse its spreading and devastating impacts on the continent.

The research describes how women may use their agency to challenge cultural institutions, although they may be aware of institutional boundaries in the process. They cannot change overnight the cultural institutions that construct gender ideology and inequalities, as inferiority and poverty actually construct their status in society, regardless of their agency. This implies the importance of considering the set of contextual factors and proxies that undermine women's susceptibility to AIDS, and to identify the specific resources needs of different categories of women, to design gender and culture sensitive development policies for a sustainable development. In this regard, I concur with Arnfred (2004: 26) that 'HIV/AIDS may be conceived as a key feminist issue'.

8.1.2 Vulnerability differentiation and resilience of Agni and migrant women

The second objective of this research was to provide insights into how AIDS impacts are experienced differently by women across cultures, and what processes shape their resilience and endurance in coping with vulnerability in their respective communities. Issues on how households and particularly women experience internal and external vulnerability have been discussed in Chapters 5 and 6. Results show the interrelatedness of internal and external vulnerability, and demonstrate that vulnerability is the outcome of an interaction of different factors and processes that are experienced differently by Agni and migrant households, and by individuals within the household. Women are faced with internal vulnerability with regard to the limited access to assets and resources in their everyday life. They also experience external vulnerability, that is the lack of access to an institutional or structural infrastructure in the form of markets, health care facilities, extension services, formal credit, and deforestation, which adds to their hardship and affects their agency to respond to stress. This research draws attention to how institutional and structural poverty emerges as the main cause of women's vulnerability, visible from the lack of investments in conservation techniques, reduced crop production, and poor access to the abovementioned infrastructure. My research shows how vulnerability to AIDS works at the interface with such adversities. Soil fertility management becomes a burden for farmers, especially for women who are in need of inputs in terms of labour and cash. In short, the study describes how women who have virtually no means of making a living (internal vulnerability) and are disadvantaged in terms of infrastructure and social services (external vulnerability), are in a situation of an increasing exposure to AIDS risks, even to the extent that some of them may engage in transactional sex as a coping strategy. So, to get access to land, extension services and credit, are crucial means not only to achieve optimal productivity for one's household food security, but also indirectly to reduce the risk of HIV or AIDS infection.

The lack of access to agricultural extension services and formal credit are considered crucial in influencing food productivity and food security at household, community and national levels. The agricultural extension policy and microfinance policy in Côte d'Ivoire has not yet shifted away from the conventional top-down approach that focuses on the male farmers, towards a more gender-based approach that particularly addresses women's needs, as they are the ones who provide the bulk of the food production (Karuhanga, 2008; FAO, 2008). Therefore, the response to AIDS requires a re-examination of Ivorian agricultural policy, along with the policy of the ANADER extension services, as AIDS-affected households will not be able to pay for extension services while they are continuously faced with the depletion of their financial resources. In this regard, my research contrasts with that of Qamar (2001), who stresses the importance of reforming the extension service provision through privatization as part of the AIDS mitigation strategies. Instead, although the ANADER has been privatized, its intervention is more focused on sensitization than on the provision of extension services. The institution does not provide free agricultural technical support as it is expected to do to the potentially AIDS-affected households.

Research findings indicate that both Agni and migrant women experience vulnerability in terms of food and livelihoods security, but in different ways. Although matrilineal Agni women have access to land, their claim to property rights of land remains problematic as land access is controlled by their brothers, while migrant women are not entitled at all to land rights. Remarkably, despite their access to land, Agni households are more food insecure than migrant households, as the Agni households are more often AIDS-affected. While food insecurity could be associated with the higher level of AIDS affection in Agni households, the interaction between AIDS, social conditions like the lack of male labour, and environmental factors, like scarcity of good soils and the lack of access to extensions services, appears to aggravate their food insecurity. Moreover, this study shows that, for Agni women, the increase of the frequency and importance of funerals in Agni culture and the related costs and time allocation aggravate the burden of the agricultural labour shortage. Consequently, food security from agricultural production in Agni households is evidently less than food security in non-farming, migrant households, who can buy food and are less dependent on the natural environment. In short, this research demonstrates that even if land is a major asset and the source of livelihood generation, in an AIDS context, when women cannot compensate for agricultural labour shortages, their access to land does not necessarily improve their socio-economic status. In fact, the reverse appears to be true, as land ownership reduces rather than yields household food security.

Studies on social capital in a context of ill heath adversity mostly state that AIDS impacts lead to an erosion of social capital, due to the lack of resources to invest in it. It is argued that, since people cannot reciprocate they cannot rely anymore on a strong social network to support their resilience to AIDS impacts and other shocks and stress (Putnam, 2000; Nombo, 2007; Nombo and Niehof, 2008). However, this research shows different results regarding social capital in a context of AIDS-related morbidity and mortality. In a situation of widespread poverty, the social obligations embodied in the social institutions in Y.F. are still being respected. Particularly the Agni show the scope of the importance of funerals in Agni culture. If an Agni fails to fulfill the social obligation *vis-à-vis* funerals to provide a financial contribution, he or she becomes exposed to exclusion from the Agni community, including the sacred connections with the ancestors. For the Agni, this is a moral death sentence they cannot cope with. In other cases social capital may be fragile, especially in reciprocating labour and material support, because the crisis affects households equally and simultaneously. Households may face difficulties to receive support from kin, neighbours and wider networks to finance medical treatment. Yet, there is a clear exception for funerals in Agni community, where people do not have a social-cultural choice but to fulfill their obligations to contribute to an increased number of funerals. In other words, this research demonstrates that social capital is more reliable in a situation of AIDS-induced mortality than in a situation of AIDS-induced morbidity.

8.1.3 Theoretical implications of women's vulnerability, agency, and resilience

The research shows how the scope of women's agency to cope with vulnerability shapes their position and value in their society. While challenging adversity in their everyday life to fulfill their gender roles as caregivers and as producers, processors, and providers of food, they show how the multilayered process of their resilience is socially and culturally embedded, and how resilience is gendered. Therefore, this research contributes to document the subjective and qualitative dimensions of resilience on the one hand, and the gendered dimensions of resilience on the other.

I have used the definition of vulnerability of Brons *et al.* (2007) on the internal vulnerability experienced at a micro and household level, when people have limited access to assets and resources, and the external vulnerability of the institutional context in terms of limited access to extension services, markets, credits, and health care, which increases household vulnerability. However, drawing on my research findings, I come to different results in terms of the way in which Agni and migrant women perceive vulnerability.

In social science and particularly in the literature on livelihoods and vulnerability, there is a tendency to split vulnerability into internal and external vulnerability, as it is perceived by individuals. When it describes the scope of women's vulnerability and the way they use their agency and resilience to deal with vulnerability, this research shows that, contrary to Brons *et al.* (2007), people themselves actually perceive a continuous relationship between internal and external vulnerabilities. Although analytically, external vulnerability increases their internal vulnerability, Agni and migrant women do not feel any differentiation between internal and external vulnerability, and they do not have separate strategies to counter it. What matters for them, is the hardship they face, which they have to challenge in their everyday life. For the Agni and migrant women, the difference rather lies in how they use their agency to cope with vulnerability, more than whether this vulnerability is internal or external to the household.

Accordingly, based on existing literature on structure and agency, and drawing on the findings of this research, I argue that Agni and migrants' social and organizational processes function as a nexus of micro and macro relations and representations (Long, 2001: 61). I argue that internal and external vulnerability are not disembodied entities. Another important finding is that, regardless of the scope of women's vulnerability, their social and gender roles as caregivers and producers, processors, and providers of food do not change. They do not have another choice than to prove that they are able to use their agency to fulfill these socio-cultural roles in any adversity context in order to be valued as good members of society. In short, internal and external vulnerability together are experienced as a common set of vulnerabilities. This implies a rejection of a unilinear or uniform concept of structural vulnerability.

The research concludes that gender-sensitive policies and programmes should be designed with regard to the needs assessment relative to the site-specific vulnerability context, and also with regard to women's ability and the scope of their agency to counter adversities, in other words, their resilience.

8.1.4 Methodological notes

This research sheds light on the complementarity of quantitative and qualitative findings with regard to households' AIDS-affection, food security, and social capital. During the survey, I came across people's reluctance to give the right information about their household's AIDS status and food insecurity. Household heads were not keen to mention whether a member was suffering from AIDS. Nor were they keen to tell about their households' food insecurity, because of the shame if the data became public. However, during in-depth interviews, FGDs, case studies and participant observation, more detailed and private kinds of information was given. During separate FGDs, women from either Agni or migrant groups did not hesitate to discuss their households' food insecurity and the overall vulnerability they faced. For example, while in the survey, households reported the impossibility to reciprocate social obligations, the data obtained through qualitative techniques complemented the survey data by revealing that, for the Agni, their vulnerability does not impact on their fulfilment of reciprocity obligations at funerals, due to the social and cultural importance of funerals and related support demands. Besides, qualitative methods have been effective to capture the meaning of sayings that illustrate food coping strategies practices, and the rationale behind women's position and status in society. The saying 'we have now become Dioulas', for example, referred to the practice of borrowing some of the cheapest food recipes from migrant women to cope with food shortages. Likewise, the saying 'the future of children reflects the submissive behaviour of their mums *vis-à-vis* their dads' referred to the experience that only women who endure the domination of their husbands can get accomplished children. In short, using both qualitative and quantitative methods complementing and sometimes contradicting each other's results permitted me to capture the dynamics of the social and agricultural impacts of AIDS, women's ability to cope with it, and the processes involved.

8.2 Policy implications and recommended approaches

The third objective of the research was to provide insights into the effectiveness of Ivorian AIDS policy and food policy on the impacts of AIDS. Drawing on the findings related to the institutional and structural conditions that emerge as the main causes of women's poverty and vulnerability, the research demonstrates that Ivorian AIDS and food policies need to be more efficient to mitigate the negative effects of AIDS on food security in the rural areas. The research sheds light on considerations and recommendations for policymakers, in several domains. In Côte d'Ivoire, these recommendations may contribute to challenge the socio-economic impacts of the recent civil war, which has affected public and private services, and may also contribute to and improve the post-conflict national development agenda. In this regard, policymakers should improve the existing agricultural and food policies, the AIDS mitigation policy and interventions, the public health policy, as well as the national environmental policy. It is obvious that the interactions between AIDS, food security, gender, livelihood, forest degradation, and context-specific cultural practices require new strategic approaches that are based on a multisectoral and integrated interventions approach.

As long as the rationale of the national agricultural policy is still heavily export-oriented, more than food security-oriented, this will continue to seriously impact any sustainable food security in Côte d'Ivoire. Food security is directly related to AIDS at a local as well as on a national level. In fact, the little interest given to food production policy not only impacts households' food security at community, regional and national levels, but it also jeopardizes rural women's livelihood generation and their socio-economic development. Therefore, policymakers should seek to develop gender-sensitive agricultural and food policies that promote food crops production, accompanied by the availability of agricultural inputs and technologies for women in order to ensure and sustain food crop production. It is obvious that in Côte d'Ivoire, the government cannot challenge the impacts of AIDS on food security without seeking to restructure and revitalize the national agricultural extension organization ANADER, both technically and financially.

Definitely, ANADER plays a key role in the prevention and mitigation of AIDS effects, in order to protect and sustain the local and national food security. ANADER has developed a policy and strategic framework to mainstream the response to AIDS in the agricultural sector. Specific activities put forward by ANADER have included awareness-raising on AIDS among the ANADER staff and rural communities, the provision of voluntary testing and counselling (VTC) services, a support programme for children made vulnerable or orphaned by AIDS, and home-based care (HBC) for

HIV-positive persons. But the lack of access to agricultural social services, such as free access to extension services, seeds and fertilizers for resource-poor and AIDS-affected households is missing in ANADER's mainstream policy, strategies and interventions. In short, their agricultural policy should address the challenges related to AIDS impacts on agriculture and food security, by reinforcing and spreading the current ANADER AIDS mainstreaming activities at the national level. These include the above-mentioned programmes, as well as the provision of financial and human resources to ANADER to strengthen its performance. Additional policy development is needed on the support of the food production and on the promotion of an educational programme on nutrition to prevent and mitigate AIDS; the provision of extension services without putting a financial burden on the farmers; providing agricultural inputs and access to formal credit to promote and strengthen income-generating activities, primarily for AIDS-affected households. These policy interventions, finally, should be available to Ivorian citizens, irrespective of whether they are of autochthonous or migrant origin.

Furthermore, it is apparent that the relevance of the national AIDS mitigation policy in Côte d'Ivoire actually lies with the inclusion rather than the exclusion of the key ministries involved in the response to the epidemic in the country. The exclusion as a result of competition between the Ministry of the Response to AIDS and the Ministry of Public Health hampers the sustainability of the response to the epidemic, and obstructs the mitigation of the AIDS-related impacts, particularly on food security. Policy makers should reduce the impacts of the political interests on mitigation policies and interventions. Additionally, they should make sure the Anti Retro Viral treatment for HIV-positive persons is available for free, without asking for a contribution (1000 FCFA) from resource-poor people. In a context of the integrated interventions required by the mitigation policy, it is imperative to promote a national social security system, which grants access to social services, transportation means and health facilities to rural resource-poor and AIDS-affected people. AIDS programmes, such as the prevention from mother-to-child transmission programme, should make formula milk available to women who cannot afford to buy it for their babies in order to prevent breast-feeding beyond the tenth month, and should provide grants for income-generating activities for HIV-positive mothers in the programme. Otherwise, the susceptibility of potentially HIV-positive children will increase.

AIDS-related cultural stigmatization definitely remains one of the biggest challenges for policymakers. Extending public education about AIDS, along with the involvement of the traditional political authorities in the AIDS mitigation process, can contribute to a reduction of AIDS-related stigmatization. Our research has shown that the cultural-historical position of power and the authority of the queen-mother in Agni society can be an asset in the fight against AIDS.

In conclusion, AIDS policies and interventions in Côte d'Ivoire should be designed to strengthen women's resilience and empowerment with the means to develop and protect their income-generating activities, in order to reduce the impact of AIDS on food security. In this regard, I agree with Nombo (2007) that the systemic nature of the epidemic lies with continued efforts to address the structural causes of poverty, with a strong emphasis on the promotion of gender equality, along with the need for dynamic, diverse and long-term responses. AIDS policies and interventions should be poverty reduction oriented by 'supporting livelihoods, strengthening food security, enhancing nutrition and health, easing the burden of women, and addressing the issue of orphans' (Nombo, 2007: 218). An efficient response to the devastating impacts of AIDS calls for advancing gender equality by efforts that especially address strategic gender needs, with an emphasis on support for vulnerable people like female heads of household, divorced women and widows (Wiegers, 2008). These efforts should include access to assets and resources, enforcing the Ivorian laws relative to ownership and inheritance rights of land for women, making financial services available and accessible for women, and designing gender-sensitive social protection. Addressing women's strategic needs in order to reverse the social and economic susceptibilities they are faced with, definitely remains the most effective strategy in reducing AIDS impacts on food security and the overall development process in Côte d'Ivoire.

8.3 Addressing the cultural challenges in the national AIDS mitigation policy

Cultural features, like matrilineal and patrilineal kinship systems that contribute to women's vulnerability to AIDS, are likely to remain a big challenge to a sustainable AIDS mitigation policy and intervention in Côte d'Ivoire. Although it is difficult to change culture by external intervention, we can influence some cultural practices. This challenge may require interventions with a strong emphasis on the participation and involvement of local actors in the process of responding to AIDS. Therefore, it is necessary to create a democratic space or arena, which will create openness and a dialogue among the actors about experimental change and, by contrast, some cultural practices that expose women to the disease. Such a dialogue should be opened between the local political authorities, especially kings and chiefs, the queen-mother, and outsider-experts, politicians, decision makers, researchers and NGOs, who would be committed to achieve the optimum goal, a society free of AIDS. Accordingly, their interactions should aim to advocate a democratic exchange of scientific and local knowledge.

Although the traditional political authorities are said to be involved in the AIDS mitigation interventions in Côte d'Ivoire, they do not have a strong presence in the mitigation process. They cannot really take action, because the traditional political authorities are subordinate to the national administrative hierarchy of the modern state. This means that there is no equity of power, even if the *sous-préfet* is supportive, that allows an Agni chief to act. In fact, an active role of the chief or the queen-mother is very much needed in this case, because of the cultural choices women are faced with. With their involvement, they may be able to provide culturally acceptable ideas, rules, and actions that may contribute to a decrease of women's socially and culturally induced vulnerability *vis-à-vis* the epidemic.

Drawing on the results of this study, I believe that given the queen-mother's power in Agni society, she can improve the knowledge and education about AIDS, and particularly the understanding of women's vulnerability to the epidemic. She can also advocate women's social and economic empowerment, which would contribute to decreasing their exposure to the disease, so that women feel secure and not stigmatized when they ask their husbands to use condoms. She can indeed enforce the cultural acknowledgement of Agni women's access to land, and she has shown to play that role even for the migrant women in her village. Even more, the queen-mother has the power and the authority to organize and empower both Agni and migrant women, as she has shown to be able to use her status to get financial credit for them, so that they can secure their livelihood, irrespective of their ethnicity.

The point I want to make is that the involvement of the cultural and political authorities, together with that of the decision makers and researchers, is crucial to the whole AIDS mitigation process, in order to strengthen and integrate place-specific or endogenous solutions to challenge the effects of AIDS. The cultural change of some practices in Agni and migrant societies is necessary, but they can be better effectuated by the active involvement of their legitimate chief and queen-mother in the decision-making process at local and supra-local levels.

8.4 Areas for future research

Finally, this thesis recommends that further studies should be done to sustain the mitigation efforts of AIDS impacts on food and livelihood security, along with the overall development process in Côte d'Ivoire.

I recommend a gender assessment of Ivorian agricultural and food policies, including land tenure, the privatization of ANADER as the national extension service that plays a key role in the response to AIDS and food security in Côte d'Ivoire, and its missions and programmes on AIDS and food security. An assessment of agricultural subsidies and their relevance in the context of AIDS is recommended as well.

Detailed comparative studies, including areas with different cultural environments, different kinship systems, different forms of livelihood generation, and different AIDS prevalence rates, may provide insights into the ways in which cultural institutions, norms and practices concur to increase women's vulnerability to AIDS, and into how this in turn interacts with women's socio-economic vulnerability in different ways, jeopardizing a sustained food security.

Much remains to be understood about how actors reconstruct socio-cultural practices that potentially increase women's exposure to AIDS, and how women themselves can actively address cultural institutional changes, in order to enable a long-term impact of AIDS mitigation policies and interventions.

8.5 Conclusion

I conclude that AIDS is not only a health problem or a social and economic problem, but a cultural and a gender problem as well. The failure to reverse the negative impacts of AIDS in sub-Saharan African countries like Côte d'Ivoire confirms the weakness of the development ideology that presently frames the AIDS approach. Because of its structural institutional approach, mainstream AIDS policy largely neglects the socio-cultural dynamics and processes that shape AIDS impacts in specific cultural contexts. Likewise, it neglects the values, motivations and perspectives of the people who have to respond to the disease. AIDS policy, particularly in the rural areas, has ignored the complex social properties of cultural value differences. Instead, the AIDS epidemic in Côte d'Ivoire requires a new approach, focusing on the existing life world of the individuals and the social groups who are exposed to the disease, especially in rural areas to correct the present urban bias. This necessitates the participation of local actors in the formulation and implementation of mitigation policies and interventions. A gender perspective adds to our understanding of the role played by women's agency and social space in their struggle to challenge their vulnerability to AIDS. In any case, the AIDS effects remain context or area specific; therefore, mitigation policies should be a response to the different cultural, social, and economic conditions.

References

Amadiume, I. (1987). *African matriarchal foundations: the Igbo case*. Karnak House: London, UK.

Amon, F.J.A. (1960). *Croyances religieuses et coutumes juridiques des Agni de la Côte d'Ivoire*. Editions La Rose: Paris, France.

Anderson, P. (1989). The myth of sustainable logging: the case for a ban on tropical timber imports. *The Ecologist* 19(5): 166-168.

Arce, A. (2003). Value contestations in development interventions: community and sustainable livelihoods approaches. *Community Development Journal* 38(3): 199-212.

Arnfred, S. (2004). *Re-thinking sexualities in Africa*. The Nordic Africa Institute: Uppsala, Sweden.

Banerjee, A. and A. Newman. (1994). Poverty, incentives and development. *American Economic Review, AEA Papers and Proceedings* 82(2): 211-215.

Barnett, T. and A. Whiteside (2002). *AIDS in the twenty-first century: disease and globalisation*. Palgrave Macmillan: Basingstoke, UK.

Bebbington, A. (1999). Capitals and capabilities: a framework for analysing peasant viability, rural livelihoods and poverty. *World Development* 27: 2021-44.

Bekele, H. (2010). *Targeting married women in microfinance programs: transforming or reinforcing gender inequalities? Evidence from Ethiopia*. Wageningen: PhD Thesis Wageningen University.

Berry, S. (1989). Social institutions and access to resources. *Africa* 59(1): 41-55.

Blaikie, P. (1989). Environment and access to resources in Africa. *Africa* 59(1): 18-40.

Blaikie, P., T. Cannon, I. Davis and B. Wisner (1994). *At risk: natural hazards, people's vulnerability and disasters*. Routledge: New York, USA.

Bleek, W. (1975). *Marriage, inheritance and witchcraft. A case study of a rural Ghanaian family*. Rijksuniversiteit Leiden, Africa Studies Centre: Leiden, the Netherlands.

Brons. J., T. Dietz, A. Niehof and K. Witsenburg (2007). Dimensions of vulnerability of livelihoods in less-favoured areas: interplay between the individual and the collective. In: R. Ruben, J. Pender and A. Kuyvenhoven (eds.) *Sustainable poverty reduction in less-favoured areas*, CABI: Wallingford, UK, pp. 91-111.

Bourdieu P. (1977). *Outline of a theory of practice*. Cambridge University Press: Cambridge, UK.

Bourdieu P. (1985). The forms of capital. In: J.G. Richardson (ed.) *Handbook of theory and research for the sociology of education*, Greenwood: New York, USA, pp. 241-258.

Bourdieu P. (1990). *The logic of practice*. Polity Press: Cambridge, UK.

Bourdieu P. (1992). *In other words: essays towards a reflexive sociology*. Polity Press: Cambridge, UK.

Bourdieu P. (1998). *Practical reason*. Polity Press: Cambridge, UK.

Bourgouin, H. and P. Guilhaume (1979). *Côte d'Ivoire 'Economie et Société'*. Editions Stock: Paris, France.

Burawoy, M. (1998). The extended case method. *Sociological Theory* 16(1): 4-33.

Bruce, J. and C.B. Llyod (1995). Finding the ties that bind: beyond headship and household. In: A. van der Avort, K. de Hoog and P. Kalle (eds.) *Single parent families*. The Netherlands Family Council: The Hague, the Netherlands, pp. 60-90.

Campbell, B.M. S. Jeffrey, W. Kozanayi, M. Luckert, M. Mutamba and C. Zindi (2002). *Household livelihoods in semi-arid regions. Options and constraints*. Center for International Policy Research: Jakarta, Indonesia.

Campilan, D. (1998). Preface: livelihood issues in rootcrop research and development. In: *Sustainable livelihood for rural households: Contribution from rootcrop agriculture*, UPWARD: Los Baños, Philippines, pp. ix-xii.

Caldwell, J.C. (1979). Education as a factor in mortality decline: an examination of Nigerian data. *Population Studies* 3 (3): 395-413.

Caldwell, J. and P. Caldwell (1990). High fertility in sub-Saharan Africa. *Scientific American* 262: 118-125.

Caldwell, J.C., P. Caldwell, E.M. Ankrah, J.K. Anarfi, D.K. Agyeman, K. Awusabo-Asare and I.O. Orubuloye (1993). African families and AIDS: context, reactions and potential interventions. *Health Transition Review* 3, Supplement: 1-16.

Carpenter, S., B. Walker, J.M Anderies, and N. Abel (2001). From metaphor to measurement: Resilience of what to what? *Ecosystems* 4: 765-781.

Carter, A.T. (1995). Agency and fertility: for an ethnography of practice. In: S. Greenhalgh (ed.) *Situating fertility: Anthropology and demographic inquiry*, Cambridge University Press: Cambridge, UK, pp. 55-86.

Chambers, R. (1990). Editorial introduction: vulnerability, coping and policy. *IDS Bulletin* 20(2): 1-7.

Chambers, R. and R. Conway (1992). *Sustainable rural livelihood: Practical concepts for the 21st century*. IDS Discussion Paper 296. Sussex: Institute of Development Studies.

Chant, S. (1997). *Women headed households: Diversity and dynamics in the developing world*. MacMillan Press: London, UK.

Cheal, D. (1989). Strategies of resource management in household economies: moral economy or political economy. In: R. Wilk (ed.) *The household economy: reconsidering the domestic mode of production*. Westview Press: Boulder, USA, pp. 11-22.

Christiaensen, L.J., R.N. Boisvert and J. Hoddinott (2000). *Validating operational food insecurity indicators against a dynamic benchmark: Evidence from Mali*. Policy Research Working Paper No. 2471. The World Bank: Washington DC, USA.

Clay, C.D. and H.K. Schwartzweller (1991). Introduction: researching household strategies. *Research in Rural Sociology and Development* 5: 1-11.

Cruz, W. and C. Gills (1990). Resource policy reform in the context of population pressure: the Philippines and Nepal. In: D. Chapman (ed.), *Arresting renewable resource degradation in the Third World*. Environment Working Paper No. 44. The World Bank: Washington DC, USA.

Davidson, J., D. Myers and M. Chakraborty (1992). No time to waste: poverty and the global environment. Oxfam: Oxford, UK.

Davis, S. (1993). Are coping strategies a cop out?. *IDS Bulletin* 24(4): 60-72.

Desai, S. and S. Alva (1998). Maternal education and child health: is there a strong causal relationship? *Demography* 35(1): 71-81.

Devereux, S. (2001). Livelihood insecurity and social protection: a re-emerging issue in rural development. *Development Policy Review* 19(4): 507-519.

De Waal, A. and J. Tumushabe (2003). *HIV/AIDS and Food Security in Africa*. DFID: London, UK.

DFID, (2000). *Sustainable livelihood guidance sheets*. Department for International Development: London, UK. Available at: at www.livelihood.org/info/info-guidancesheets.htm (12.04.07).

Dietz, T. (2000). *Poverty and land degradation*. Paper for the conference on World Poverty: A challenge for the private sector. Amsterdam, the Netherlands, October 3-4, 2000.

Dietz, A.J., A. Verhagen and R. Ruben (eds.) (2004). *The impact of climate change on dry lands. With a focus on West Africa.* Kluwer Academic Publishers: Dordrecht, the Netherlands.

Donahue, J., K. Kubbucho and S. Osinde (2001). *Responding to a silent economic crisis among microfinance clients in Kenya and Uganda.* Available at: http://www.microsave.com (accessed 12.05.08).

Duraiappah, A.K. (1996). *Poverty and environmental degradation: a literature review and analysis.* GREED Working Papers Series No 8. IIED: London, UK.

Douglas, M. (1969). Is matriliny doomed in Africa? In: M. Douglas and P.M. Kaberry (eds.) *Man in Africa.* Tavistock: London, UK, pp. 121-135.

DSRP, (2009). *Stratégies de relance du développement et de réduction de la pauvreté.* Ministère de la Planification: Abidjan, Côte d'Ivoire.

Duchesne, V. (1994). Lier, délier. Les amwan chez les Anyi de Côte-d'Ivoire. *Journal des Africanistes* 64(2): 129-139.

EIS (2005). *Enquête sur les indicateurs du Sida en Côte d'Ivoire.* Ministère de la Lutte contre le Sida, Abidjan, Côte d'Ivoire.

Ellis, F. (2000). *Rural livelihoods and diversity in developing countries.* Oxford University Press : Oxford and New York.

ENV (2008). *Enquêtes sur le niveau de vie des ménages en Côte d'Ivoire.* Institut National de Statistiques: Abidjan, Côte d'Ivoire.

Eschlimann, J.P. (1985). *Les Agni devant la mort (Côte d'Ivoire).* Karthala: Paris, France.

Fairhead J. and M. Leach (1998). *Reframing forest degradation: global analysis and local realities: studies in West Africa.* Routledge: London, UK.

FAO (1996). *Rome: declaration on world food security,* Proceedings from the World Food Summit, November 13-17, 1996. Rome, Italy. Available at: http://www.fao.org (accessed 15.3.07).

FAO (1999). *Implications of economic policy for food security: a training manual.* FAO: Rome, Italy.

FAO (2000). *Tropical forest degradation literature: geographical and historical patterns in the availability of information and the analysis of causes.* FAO: Rome, Italy.

FAO (2004). *HIV/AIDS and the food crisis in sub-Saharan Africa: the impacts of HIV/AIDS on food security,* Twenty-third Regional Conference for Africa, 1-5 March 2004, Johannesburg, South Africa.

Fieloux, M. (1980). *Les migrations rurales des Lobis de la Haute Volta vers la Côte d'Ivoire.* ORSTOM: Paris, France.

Firebaugh, F.M. (1994). Female-headed households: survival strategies and food security. In: K. de Hoog and J.A.C van Ophem (eds.) *Changes in daily life,* Wageningen Agricultural University: Wageningen, the Netherlands, pp. 89-99.

Fox, R. (1967). *Kinship and marriage: an anthropological perspective.* Penguin: Harmondsworth UK.

Frankenberger, T.R., and M.K. McCaston (2001). *The household livelihood security concept.* Available at: http://www.fao.org/docrep/x0051t05.htm (accessed 05.01.06).

Fraser, E.D.G. (2003). Social vulnerability and ecological fragility: building bridges between social and natural sciences using the Irish Potato Famine as a case study. *Conservation Ecology* 7(2): 9.

Garret, L. (2007). The challenge of global health. *Foreign affairs*: 86: 1-17.

Glewwe, P. and G. Hall (1998). Are some groups more vulnerable to macro-economic shocks than others? Hypothesis tests based on panel data from Peru. *Journal of Development Economics* 56: 181-206.

Giddens, A. (1984). *The Constitution of society: outline of the theory of structuration.* University of California Press: Berkeley, USA.

Giddens, A. (1979). *Central problems in social theory: action, structure and contradiction in social analysis.* Macmillan Press: London, UK.

Gillespie, S. and S. Kadiyala (2005). HIV/AIDS and food and nutrition security: From evidence to action. *Food Policy Review* 7: 195-208.

Green, J.C., V.J. Caracelli and W.D Graham (1989). Toward a conceptual framework for mixed-method evaluation designs. *Educational Evaluation and Policy Analysis* 11(3): 255-274.

Gupta, G.R. (2000). *Vulnerability and resilience: gender and HIV/AIDS in Latin America and the Caribbean.* ICRW: Washington DC, USA.

Guyer, J.I. and P.E. Peters (1987). Conceptualizing the household: issues of theory and policy in Africa. *Development and Change* 18(2): 1197-1214.

Haddad, L. and S. Gillespie (2001). *Effective food and nutrition policy responses to HIV/AIDS: what we know and what we need to know.* Discussion Paper No 112. IFPRI: Washington DC, USA.

Haggblade, S., P. Hazell and T. Reardon (2002). *Strategies for stimulating poverty-alleviating growth in the rural non-farm economy in developing countries.* EPTD Discussion Paper No.92. IFPRI: Washington DC, USA.

Hartmann, B. (2001). Will the circle be unbroken? A critique of the project on environment, population, and security. In: N.L. Peluso and M. Watts (eds.). *Violent Environments.* Cornell University Press: Ithaca, USA, pp. 39-62.

Hill, P. (1963). *The migrant cocoa farmers of Southern Ghana; a study in rural capitalism.* Cambridge University Press: London, UK.

Homer-Dixon, T.F. (1999). *Environment, scarcity, and violence.* Princeton University Press: Princeton, USA.

Hulme, D. and P. Mosley (1996). *Finance against poverty.* Volume 1, Routledge: London, UK.

Huq, H. (2000). *People's practices. Exploring contestation, counter-development, and rural livelihoods. Cases from Muktinagar, Bangladesh.* PhD. Thesis Wageningen University: Wageningen, the Netherlands.

IFAD (2001). *Gender and households.* Rome: International Fund for Agricultural Development. Available at: http://www.ifad.org/gender/approach/hfs/wfs_3.htm (accessed 15-4-10).

INS (2008). *L'Enquête sur le niveau de vie des ménages en Côte d'Ivoire.* Institut National de Statistiques: Abidjan, Côte d'Ivoire.

Kabeer, N. (1991). *Gender, production and well-being: Rethinking the household economy.* Discussion paper 228. Institute of Development Studies: Sussex, UK.

Kabeer, N. (1995). *Necessary, sufficient, or irrelevant: women, wages, and intrahousehold power relations in urban Bangladesh.* Working Paper No. 25, Institute of Development Studies: Sussex, UK.

Karuhanga, B.M. (2008). *Living with AIDS in Uganda: impacts on banana farming households in two districts.* AWLAE series 6. Wageningen Academic Publishers: Wageningen, the Netherlands.

Kessy, F. and B. Obrist (2008). Identifying resilience pathways to sanitary health problems in an unplanned ward of Dodoma, Tanzania. *Medische Antropologie* 20 (2): 227-240.

Kofi, A.A. (1990). Matriliny and the new Interstate Succession Law of Ghana. *Canadian Journal of African Studies / Revue Canadienne des Études Africaines* 24(1) : 1-16.

Krishna, A. (2004). Escaping poverty and becoming poor: who gains, who loses, and why? *World Development* 32: 121-136.

Knorr-Cetina, K. and A. Cicourel (eds.) (1981). *Advances in social theory and methodology: toward an integration of micro- and macro-sociologies*. Routledge & Kegan Paul: Boston, USA.

Koster, M. (2009). *In fear of abandonment: slum life, community leaders and politics in Recife, Brazil.* PhD Thesis Wageningen University: Wageningen, the Netherlands.

Leach, M. and J. Fairhead (2000). Fashioned forest pasts, occluded histories? International environmental analysis in West African locales. In: Doornbos, M., A. Saith and B. White (eds.), *Forests, nature, people, power*. Blackwell Publishers: Oxford, UK, pp. 35-59.

Lévi-Strauss, C. (1963). *Structural anthropology.* Basic Books: New York, USA.

Loevinsohn, M. and S. Gillespie (2003). *HIV/AIDS, food security, and rural livelihoods: understanding and responding*. Regional Network on HIV/AIDS, Rural Livelihoods and Food Security. Working Paper No. 2. RENEWAL: Washington DC, USA.

Long, N. (1989). *Encounters at the interface: a perspective on social discontinuities in rural development.* Wageningen Agricultural University: Wageningen, the Netherlands.

Long, N. (2001). *Development sociology: Actors perspectives.* Routledge: London, UK.

Long, N. and A. Long (1992). *Battlefields of knowledge: the interlocking of theory and practice in social research and development.* Routledge: London, UK.

Long, N. and J.D. van der Ploeg (1989). Demythologizing planned intervention: an actor perspective. *Sociologia Ruralis* XXIX(3/4): 226-249.

Long, N. and J.D. van der Ploeg (1994). Heterogeneity, actor, and structure: towards a reconstitution of the concept of structure. In: D. Booth (ed.) *Rethinking social development: theory, research and practice.* Longman: Harlow, UK, pp. 62-89.

Long, N. and J.D. van der Ploeg (1995). Reflections on agency, ordering the future and planning. In: G.E. Frerks and J.H.B. den Ouden (eds.) *In search of the middle ground: essays on the sociology of planned development*, Department of Sociology, Wageningen Agricultural University: Wageningen, the Netherlands, pp. 64-78.

Loukou, J.N. (1977). *La Reine Pokou: Fondatrice du royaume Baoulé (Grandes Figures Africaines).* Nouvelles Editions Africaines: Abidjan, Côte d'Ivoire.

Machera, M. (2004). Opening a can of worms: a debate on female sexuality in the lecture theatre. In: S. Arnfred (ed.) *Re-thinking sexualities in Africa.* Nordic Africa Institute: Uppsala, Sweden, pp. 157-172.

Matinga, M.N. (2008). The making of hardiness in women's experience of health impacts of wood collection and use in Cuntwini, rural South Africa. *Medische Antropologie* 20(2): 279-296.

Maxwell, S., and T.R. Frankenberger (1992). *Household and food security: concepts, indicators, measurements.* UNICEF and IFAD: New York, USA.

Meyer, B. (1995). *Translating the devil. An African appropriation of pietist protestantism. The case of the Peki Ewe in Southeastern Ghana, 1847-1992.* Ph.D. Thesis, University of Amsterdam: Amsterdam, the Netherlands.

Ministère de la Lutte contre le SIDA (2005). *L'Enquête sur les indicateurs du Sida en Côte d'Ivoire.* Abidjan, Côte d'Ivoire.

Morduch, J. (1994). Poverty and vulnerability. *American Economic Review, Papers and Proceedings* 84(2): 221-225.

Moser, C.O.N. (1993). *Gender planning and development: Theory, practice and training.* Routledge: London, UK.

Moser, C.O.N. (1996). *Confronting crisis: a comparative study of household responses to poverty and vulnerability in four poor urban communities*. Environmentally Sustainable Development Studies and Monographs Series No. 8. The World Bank: Washington DC, USA.

Moser, C.O.N. (1998). The asset vulnerability framework: reassessing urban poverty reduction strategies. *World Development* 26(1): 1-19.

Müller, T.R. (2004). *HIV/AIDS and agriculture in sub-Saharan Africa*. AWLAE Series 1. Wageningen Academic Publishers: Wageningen, the Netherlands.

Müller, T.R. (2005a). *HIV/AIDS, gender and rural livelihoods in sub-Saharan Africa*. AWLAE Series 2. Wageningen Academic Publishers: Wageningen, the Netherlands.

Müller, T.R. (2005b.). *HIV/AIDS and human development in sub-Saharan Africa*. AWLAE Series 3. Wageningen Academic Publishers: Wageningen, the Netherlands.

Murray, C. (2000). Livelihood research: Transcending boundaries of time and space. *Journal of Southern African studies* 28(3): 489-509.

Mutangadura, G. (2000). *Household welfare impacts of mortality of adult females in Zimbabwe: implications for policy and program development*. Paper presented at The Aids and Economics Symposium, organized by the International AIDS Economics Network (IAEN). Durban, South Africa, 7-8 July 2000.

Nauta, W. (2008). Saving the deprived Africans: how the fight against HIV/AIDS has cast the victims. Paper presented at the Conference on The Ends of Development: Market, morality, religion, political theology? VU University, Amsterdam, the Netherlands, 19-20 June 2008.

Narayan, D., R. Patel, K. Schafft, A. Rademacher and S. Koch-Schulte (2000). *Voices of the poor. Can anyone hear us?* Oxford University Press for the World Bank: New York and Washington DC, USA.

Negash, A. (2001). *Diversity and conservation of enset and its relation to household food and livelihood security in South-western Ethiopia*. PhD. Thesis, Wageningen University: Wageningen, the Netherlands.

Netting, McC.R. and R.R. Wilk (1984). Households: changing forms and functions. In: McC.R. Netting, R.R. Wilk, and E.J. Arnould (eds.) *Households, comparative and historical studies of the domestic group*. The University of California Press: Berkeley, USA, pp. 1-28.

Nguthi, F.N. (2007). *Adoption of agricultural innovations by smallholder farmers in the context of HIV/AIDS: the case of the tissue cultured banana in Kenya*. AWLAE series 7. Wageningen Academic Publishers: Wageningen, the Netherlands.

Niehof, A. (1999). The gendered nature of household food security. In: A. Niehof and P. Tepstra (eds.), *Household in an interdisciplinary perspective. Liber Amicorum for Antine Hardon-Baars*, Household and Consumer Studies Publication Series 2, Wageningen Agricultural University: Wageningen, the Netherlands, pp. 105-119.

Niehof, A., and L. Price (2001). *Rural livelihood systems: a conceptual framework*. Wageningen-UPWARD Series on Rural Livelihoods No.1. WU-UPWARD: Wageningen, the Netherlands.

Niehof, A. (2003a). Why kinship, what is it and how to do it? Unpublished lecture notes. Wageningen University: Wageningen, the Netherlands.

Niehof, A. (2003b). Gendered dynamics of food security at the household level. In: K. Ulmer (ed.) *No security without food security-No food security without gender equality*. APRODEV: Brussels, Belgium, pp. 60-68.

Niehof, A. (2004a). The significance of diversification for rural livelihood systems. *Food Policy* 29: 321-338.

Niehof, A. (2004b). A micro-ecological approach to home care for AIDS patients. *Medische Antropologie* 16(2): 245-265.

Niehof, A. (2008). Introduction: dimensions of resilience in a context of health-related adversity. *Medische Antropologie* 20(2): 217-226.

Niehof, A. (2010). *Food, diversity, vulnerability and social change: research findings from insular Southeast Asia*. Mansholt Publication Series No. 9. Wageningen Academic Publishers: Wageningen, the Netherlands.

Niehof, A., and L. Price (2008). Etic and emic perspectives on HIV/AIDS impacts on rural livelihoods and agricultural practice in sub-Saharan Africa. *NJAS Wageningen Journal of Life Sciences* 56(3): 139-153.

Nombo, C. (2007). *When AIDS meets poverty: Implications for social capital in a village in Tanzania*. Awlae series 5. Wageningen Academic Publishers: Wageningen, the Netherlands.

Nombo, C. and A. Niehof (2008). Resilience of HIV/AIDS-affected households in a village in Tanzania: Does social capital help? *Medische Antropologie* 20(2): 241-258.

Otsuki, K. (2007). *Paradise in a Brazil nut cemetery: sustainability discourses and social action in Para, the Brazilian Amazon*. PhD Thesis Wageningen University: Wageningen, the Netherlands .

Orubuloye. I., J. Caldwell and P. Caldwell (1993). African women's control over their sexuality in an era of AIDS. *Social Science and Medicine* 37: 859-872.

Pennartz, P., and A. Niehof (1999). *The domestic domain: chances choices and strategies of family households*. Ashgate: Aldershot, UK.

Perrot. C. H. (1981). Traditions orales et archéologie: Une migration Anyi (Côte d'Ivoire). In: *2000 Ans d'histoire Africaine: Le sol, la parole et l'écrit*. Harmattan: Paris, France, 307-319.

Population Reference Bureau (2008). *World population data sheets*. Washington DC, USA.

Portes, A. (1998). Social capital: its origins and applications in modern sociology. *Annual Review of Sociology* 24: 1-24.

Putnam, R. (2000). *Bowling alone: the collapse and revival of American community*. Simon and Schuster: New York, USA.

Pritchett, L., A. Suryahadi, and S. Sumarto (2000). *Qualifying vulnerability to poverty: a proposed measure, applied to Indonesia*. Policy Research Working Paper No. 2437. The World Bank: Washington DC, USA.

Qamar, M.K. (2001). The HIV/AIDS epidemic: an unusual challenge to agricultural extension services in sub-Saharan Africa. *Journal of Agricultural Education and Extension* 8(1): 1-11.

Quisumbing, A.R, L. Haddad, and R. Meinzen-Dick (1998). *Gender issues for food security in developing countries: implications for project design and implementation*. IFPRI, The World Bank: Washington DC, USA.

Radstake, M. (2000). *Secrecy and ambiguity. Home care for people living with HIV/AIDS in Ghana*. ASC Research Report 59. African Studies Centre: Leiden, the Netherlands.

Rattray, R.S. (1969). *Ashanti law and constitution*. Clarendon Press: Oxford, UK.

Rattray, R.S. (1959). *Religion and art in Ashanti*. Clarendon Press: Oxford, UK.

Redclift, M. (1990). The role of agriculture in sustainable development. In: P. Lowe, T. Marsden and S. Whatmore (eds.) *Technological change and the rural environment*, David Fulton Publishers: London, UK, pp. 81-103.

Repetto, R. (1990). Forest degradation in the tropics. *Scientific American* 262(4): 36-45.

Richards, A.I. (1950). Some types of family structures amongst the central Bantu. In: C.D. Forde and A.R. Radcliffe-Brown (eds.) *African systems of kinship and marriage*, Oxford University Press: London, UK, pp. 166-206.

Ritchie, J. (2003). The applications of qualitative methods to social research. In: J. Ritchie and J. Lewis (eds), *Qualitative research practice, A guide for social science students and researchers*, Sage Publications: London, UK, pp. 24-46.

Rudie, I. (1995). The significance of eating: cooperation, support and reputation in Kelantan Malay households. In: K.J. Wazir (ed.), *'Male' and 'female' in developing Southeast Asia*. Berg Publishers: Washington DC, USA, pp. 227-247.

Rugalema, G. (1999a). *Adult mortality as entitlement failure: AIDS and the crisis of rural livelihoods in a Tanzanian village*. Institute of Social Studies, PhD Thesis: The Hague, the Netherlands.

Rugalema, G. (1999b). It is not only the loss of labour: HIV/AIDS, loss of household assets and household livelihood in Bukoba District, Tanzania. In: G. Mutagandura, H. Jackson and D. Mukurazita (eds.) *AIDS and African smallholder agriculture*. Southern African AIDS Information Dissemination Service: Harare, Zimbabwe, pp. 41-52.

Sen, A.K. (1981). *Poverty and famines. An essay on entitlement and deprivation*. Clarendon Press: Oxford, UK.

Sen, A.K. (1990). Gender and cooperative conflicts. In: I. Tinker (ed.) *Persistent inequalities: women and world development*. Oxford University Press: Oxford, UK, pp. 123-149.

Sen, B. (2002). *Risks, vulnerability and poverty in Bangladesh: issues and evidence: variations on a theme*. Available at: http://www.bids.org/Sen2002.doc (accessed 06.05.08).

Seeley. J., K. Sutherland, R. Dey and R. Grellier (2003). *Mitigation of gender-specific impact of HIV/AIDS on rural livelihoods*. Overseas Development Group: Norwich, UK.

Somanathan, E. (1991). Forest degradation, property rights and incentives in Central Himalayas. *Economic and Political Weekly* 26: 37-46.

Scrimshaw S.C.M. (1990). Combining quantitative and qualitative methods in the study of intra-household resource allocation. In: B.L. Rogers and N.P. Schlossman (eds.) *Intra-household Resource Allocation*. United Nations University Press: Tokyo, Japan, pp. 186-198.

Schultz, T.P. (1993). Return to women's education. In: E. King and M.A Hill (eds.) *Women's education in developing countries: barriers, benefits and policies*. The John Hopkins University Press: Baltimore, USA, pp. 51-99.

Schoepf, B.G. (2001). International AIDS research in anthropology: taking a critical perspective on the thesis. *Annual Review of Anthropology* 30: 335-361.

Smith, L., U. Ramakhrishnan, A. Ndiaye, L. Haddad and R. Martorell (2003). The importance of women's status for child nutrition in developing countries. In: A.R. Quisumbing (ed.), *Household and decisions, gender and development: a synthesis of recent research*. The John Hopkins University Press for IFPRI: Baltimore, USA, pp. 41-50.

Smith, S. and D. Cohen (2000). *Gender, development and the HIV epidemic*. Available at: http://www.undp.org/hiv/publications/gender/gendere.htm (accessed 22-06-08).

Southgate, D. (1990). The causes of land degradation along spontaneously expanding agricultural frontiers in the Third World. *Land Economics* 66(1): 93-101.

Spring, A. (2000). Commercialisation and women farmers in Kenya. In: A. Spring (ed.), *Women farmers and commercial ventures: Increasing food security in developing countries*, Lynne Rienner Publishers: Boulder, USA, pp. 317-341.

Stokes, S. (2003). *Measuring impacts of HIV/AIDS on rural livelihoods and food security*. Rome: FAO, Italy. Available at: http://www.fao.org/sd/2003/PE0102a_en.htm (accessed 16.11.09).

Stillwaggon, E. (2006). *AIDS and the ecology of poverty*. Oxford University Press: New York, USA.

Swift, J. (1989). Why are rural people vulnerable to famine? In: R. Chambers (ed.), Vulnerability, coping and policy. *IDS Bulletin* 20(2): 8-15.

Tapouzis, D. and J. du Guerny. (1999). *Sustainable agriculture: rural development and vulnerability to the AIDS epidemic.* UNAIDS Best Practice Collection: UNAIDS and FAO: Rome, Italy.

Tapouzis, D. (2000). *Measuring the impact of HIV/AIDS on the agricultural sector in Africa.* UNAIDS Best Practice Collection. UNAIDS and FAO: Rome, Italy.

UN (2000). Security Council Adopting Historic Resolution 1308 (2000) on HIV/AIDS. Available at: http://www.un.org/News/Press/docs/2000/20000717.sc6890.doc.html (accessed 10.02.07).

UNICEF, (2008). *The state of the world's children 2008: social and economical statistics on women and children of Côte d'Ivoire.* UNICEF: Geneva, Switzerland.

UNAIDS (2008a). *Report on the global AIDS epidemic*. UNAIDS: Geneva, Switzerland.

UNAIDS (2008b). Second independence evaluation 2002-2008: Country visit to Côte d'Ivoire. UNAIDS: Geneva, Switzerland.

UNDP (2007). *Human Development Report*. UNDP: New York, USA.

Van der Geest, S. (2008). Resilience and the whims of reciprocity in old age: an example from Ghana. *Medische Antropologie* 20(2): 297-313.

Van den Breemer, J.P.M. (1984). *Onze aarde houdt niet van rijst. Een cultureel antropologische studie van innovatie in de landbouw bij de Aouan van Ivoorkust.* PhD Thesis Leiden University: Leiden, the Netherlands.

Van den Top, G.M. (1998). *The social dynamics of forest degradation in the Sierra Madre, Philippines.* PhD Thesis Leiden University: Leiden, the Netherlands.

Walker, B., C.S. Holling, S.R. Carpenter, and A. Kinzig (2004). Resilience, adaptability and transformability in social-ecological systems. *Ecology and Society* 9(2): 5. Available at: http://www.ecologyandsociety.org/vol19/iss2/art5/ (accessed 12.6.09).

Westoby, J. (1987). *The purpose of forests: follies of development*. Blackwell: Oxford, UK.

Wiegers, E.S. (2008). *Gendered vulnerability to AIDS and its research implications.* PhD Thesis Wageningen University: Wageningen, the Netherlands.

Wisner, B. (1993). Disaster vulnerability: Geographical scale and external reality. In: H. Bhole (ed.) *Worlds of pain and hunger: geographical perspectives on disaster vulnerability and food security*, Verlag Breitenbach: Saarbrucken, Germany, pp. 13-55.

Wolf, E.R (1990). Facing power: old insights – new questions. *American Anthropologist* 92(3): 586-596.

World Bank (1993). *World Development Report, 1993: investing in health.* Oxford University Press: New York, USA.

Websites

http://www.infoplease.com/country/profiles/cote-d-ivoire.html (accessed 20.07.09).
http://www.go.worldbank.org/SN2JJ08P10 (accessed 24.07.09).
http://www.pepfar.gov/press/81552.htm (accessed 01.2.08).
http://fr.wikipedia.org/wiki/Abengourou (accessed 5.4.10).

Appendix

Table A1. Mixed methods used to collect data.

Design	Instrument[1]	Emphasis	Mode of administration	Contribution to the study
Reconnaissance survey	Collection of secondary data, discussions with ANADER, the principal doctors of health centres and key farmers	Exploration of 3 study areas (Y.F., Zenou, Dabe) Y.F. was finally selected to be the study area	General discussions with key informants	Verifying key issues necessary for conceptualizing the proposal, establishing contacts with key informants on the ground
Key informants	Semi-structured interview with local leaders, ANADER, the MOA, MORA, MOPH, SODEFOR officers	Qualitative data	In-depth interviews	Knowledge about the area, its people, their livelihood activities, the gender relations in labor, resource allocation, sexuality, forest degradation, AIDS and impacts in the area
Focus-Groups Discussions	Group discussions with: The chief and his notables about social structure, Agni's culture/matrilineal system, gender link livelihoods, sexuality, health/AIDS, forest degradation and food security problem The migrants male about their culture/patrilineal system, cultural cohabitation, gender link livelihoods, sexuality, health/AIDS, forest degradation and food security problem Agni and migrants Women about gender relations regarding livelihoods activities, resource allocation, cultural systems, sexuality, health/AIDS, forest degradation and food security problem	Qualitative data	Open-ended set of questions for clarification and opening out discussion to widen the range of responses	To identify key issues, perceptions, attitudes expectation, and mapping interview schedules for the household survey questionnaire

Table A1. Continued.

Design	Instrument[1]	Emphasis	Mode of administration	Contribution to the study
Survey	Structured interviews (191 farming households).	Quantitative data on household structure/ characteristics, assets and resources, intra-household allocation, household activities and outcomes, crop production, AIDS impacts and effect of forest degradation on household assets/ livelihood activities and food security	The questionnaires were administered face-to-face	To answer descriptive questions, enable generalisation, prediction leading to opportunity for further research
In-depth interviews	Semi-structured interview guide using a list of questions and topics based on the outcome of the quantitative survey and the focus groups discussions	Qualitative data allowing to understand the gender division of labor, access to and control of household assets, impacts of AIDS and effects of forest degradation on households' food and livelihood security	Validate and clarify the information collected in the survey and in the FGDs	To validate the quantitative data; to stress the extent to which AIDS impacts and effects of forest degradation on households' assets and food and livelihood security

[1] MOA: Ministry of Agriculture; MORA: Ministry of the Response to AIDS; MOPH: Ministry of Public Health; SODEFOR: National Forestry Agency; ANADER: Agence Nationale de Developpement Rural (National Agency for Rural Development).

Table A2. Mixed methods used to collect data and topics discussed.

Method	Data collection	Respondents	Topics discussed
Reconnaissance survey	Analysis of relevant reports and documents/local and international literature from researcher, journalist, activists and civil servants		AIDS situation related gender and food security Ethnographic, physical and demographic data of the study area, crop production and farming systems
	Discussions with ANADER extension officers, health officers and key farmers	ANADER extension officers, the principal doctors of health centres in visited areas	General information about health and AIDS situation and policy, related programs, relations with the national and international stakeholders Information on gender and food security Programs outcome and constraints Recommendations
		Men and women farmers in three villages	Livelihoods activities and constraints Main problems related crop production and farming systems, and food security Health and AIDS problems with emphasis on gender and food security AIDS programs in the areas Their impacts on gender and food security
Focus-Groups Discussions	FGD 1	Agni men	Community resources, livelihoods activities and strategies Social structure, Agni's culture and matrilineal system Gender link livelihoods, sexuality, health/AIDS, forest degradation and food security problem
	FGD 2	Migrant men	Community resources, livelihoods activities and strategies Social organization, land ownership and integration issues Migrant' culture and patrilineal system Gender link livelihoods, sexuality, health/AIDS, forest degradation and food security problem

Table A2. Continued.

Method	Data collection	Respondents	Topics discussed
Focus-Groups Discussions (continued)	FGD 3	Agni women	Gender relations regarding livelihoods activities, resource allocation, labor allocation cultural kinship matrilineal and patrilineal systems, sexuality, health/AIDS problems and programs, forest degradation and food secrity problems Coping strategies to overcome vulnerability Communities integration Needs assessment and challenges Recommendations to improve their situations/ empowerment
	FGD 4	Migrant women	Gender relations regarding livelihoods activities, resource allocation, labor allocation cultural kinship matrilineal and patrilineal systems, sexuality, health/AIDS problems and programs, forest degradation and food security problems Coping strategies to overcome vulnerability Communities integration Needs assessment and challenges Recommendations to improve their situations/ empowerment
	FGD 5	Agni and migrant youths	Main problems of the village Community resources and youth livelihoods activities and strategies Relations with the first generation AIDS awareness Health/AIDS, forest degradation and the food security situation with emphasis on gender issues Youth input to respond to AIDS in the village Assessment of AIDS programs in the village: outcome, constraints and recommendation to improve the mitigation programs

Survey	Structured interviews	191 farming households	Household characteristics and structure Household assets and resources, Household activities and income, crop production and consumption Effects of illness/AIDS and death Forest degradation on household assets/livelihood activities and food security
In-depth interviews	Semi-structured interview guide using a list of questions and topics based on the outcome of the quantitative survey and the focus groups discussions	ANADER AIDS program leader	Agricultural production and food security issues at national level Extension services and the beneficiaries with particular emphasis on gender issues Effects of illness/AIDS on livelihood and food security ANADER AIDS program goals and activities Strengths and weaknesses Recommendations for improving the program's outcomes
		ANADER-Abengourou office leader	Agricultural production and food security issues at the district level Extension services and the beneficiaries with particular emphasis on gender issues Effects of illness/AIDS on livelihood and food security ANADER AIDS programs implementation Strengths and weaknesses Recommendations for improving the program's outcomes
		ANADER village extension officer	Agricultural production in the village and the main constraints, with particular emphasis on gender and food security issues Extension services performance, cost, gender issues Extension services constraints, and recommendations to improve the outcome

Table A2. Continued.

Method	Data collection	Respondents	Topics discussed
In-depth interviews (continued)	Semi-structured interview guide using a list of questions and topics based on the outcome of the quantitative survey and the focus groups discussions (continued)	MOA food security program leader	National agricultural policy and food policy Food security situation in Côte d'Ivoire link with AIDS Mitigation programs Relations with stakeholders Programs' outcome, constraints and recommendations
		MORA technical director	AIDS situation in Côte d'Ivoire, with particular emphasis in rural areas and gender issues The national mitigation policy Strength and weaknesses The relations between MORA and MOPH The leadership problem and its impacts on the response to the epidemic in the country The relations with AIDS international agencies Recommendations to improve the national response to AIDS
		MOPH	AIDS situation in Côte d'Ivoire, with particular emphasis in rural areas and gender issues The national mitigation policy Strength and weaknesses The relations between the MOPH and the national and international stakeholders The relations between MORA and MOPH The leadership problem and its impacts on the response to the epidemic in the country The relations with AIDS international agencies Recommendations to improve the national response to AIDS

The doctor and district director of the anti-tuberculosis and AIDS testing center	Health/tuberculosis/AIDS problems, AIDS awareness and programs at the district level Relations with national, international, and local stakeholders Relations with the traditional chieftaincy AIDS link gender and food security in the area Mitigation programs, outcome and constraints Recommendations
SODEFOR regional director	The region ecological background Forest degradation problem, causes, consequences and actions undertaken to mitigate the problem Forest degradation link food security and gender The illegal logging activities questions Relations with local communities Outcomes, constraints and recommendations to improve the fight against forest degradation
Leader of CBO for forests protection	Ecological background of the village Forest degradation problem, causes, consequences and actions undertaken to mitigate the problem, at the community level Forest degradation link food security and gender The illegal logging activities questions Relations with the forestry officers The local ecological knowledge to protect forest Outcomes, constraints and recommendations to improve the local mitigation actions
Village chief/ 'porte-canne'	The village economic, socio-cultural, and ecological background The traditional political power and the relations with the modern state power Illness/AIDS related food security with emphasis on gender issues Migrant integration The chief input/capabilities for reversing some cultural practices exposing women to HIV/AIDS

Table A2. Continued.

Method	Data collection	Respondents	Topics discussed
In-depth interviews (continued)	Semi-structured interview guide using a list of questions and topics based on the outcome of the quantitative survey and the focus groups discussions (continued)	Queen-mother	Life history Women's status, situation in Agni culture and in the village Queen-mother power and women's empowerment Input/ capabilities of the queen-mother to impact on women's socio-cultural, and economic, vulnerability Global recommendations for women empowerment
		The *komian* (traditional healer)	Life history Agni anthropological and cultural meaning of illness and AIDS The village main health problems, links with AIDS and food security, with emphasis on women and children The relations with the hospital practitioners The *komian* and the AIDS programs in the village Input and capabilities of the *komian* to respond efficiently to AIDS in the village
		Sous-prefet	The village economic, socio-cultural, and ecological background The traditional political power and the relations with the modern state power Illness/AIDS related food security with emphasis on gender issues Migrant communities' integration Involvement of the 'sous-prefet' in AIDS programs in the village Outcome, constraints and recommendations to improve the mitigation programs Capabilities of the 'sous-prefet' to impact on some socio-cultural practices exposing women to AIDS

Leader of the village committee of the response to AIDS	AIDS awareness and situation in the village, with emphasis on gender and food security AIDS programs in the village The committe members' work and the relations with the stakeholders Outcome, constraints and recommendations to improve the response to AIDS in the village
The principal doctor of the local health center	The main health problems in the setting with emphasis on women and children Health and AIDS related food security with emphasis on women and children AIDS programs monitor by the hospital The relations with local stakeholders and other AIDS programs in the village Outcome, constraints and recommendations to improve the response to AIDS in the village
Midwife	Women main health problems with emphasis on reproduction health, AIDS and food security Assessment of the Prevention of mother-to-child transmission program Outcome, constraints and recommendations to improve the response to AIDS in the village
Traditional midwife	Background of the traditional midwife, and the relations with the hospital workers, and the MOPH Their performance in the village with emphasis on AIDS Outcome, constraints and recommendations to improve their performance
Leader of the microfinance institution (CMEC) operating in the village	Background of the CMEC, link gender and food security CMEC assessment, CMEC and AIDS problem and programs in the village CMEC and women empowerment Outcome, challenges, constraints, future actions and recommendations for women empowerment

Summary

The AIDS epidemic is still challenging the world-wide efforts to be reversed in sub-Saharan African countries. There is a consensus that the region remains the most affected, and women the most susceptible to the epidemic, regardless of the large amounts of human and financial resources allocated to respond to the disease. This study questions whether the mainstream policies and interventions to respond to AIDS in countries like Côte d'Ivoire are effective, as they are based on a dominant biomedical ideology and the neglect of the socio-cultural dynamics that influence women's vulnerability to the disease, and its impact on food and livelihood security.

This study investigates how matrilineal and patrilineal kinship and differential cultural norms and practices increase women's exposure to AIDS, and how these influence the way they respond to it by using their agency and capabilities to cope with AIDS and socio-economic hardship, along with other adversities, such as forest degradation.

The study was carried out in Y.F., a sub-district of Abengourou, the second most AIDS-affected district of Côte d'Ivoire. Y.F. is an important area in terms of food production, and it is a multicultural setting as well. The majority consists of autochthonous Agni landowners, but migrants from other regions of Côte d'Ivoire, as well from neighbouring countries like Ghana, Burkina, Mali, Guinea, Niger and Nigeria, have settled in the region since the early 1970s.

The study was designed by a combination of quantitative and qualitative methods. A survey of 191 households was supplemented by a month-long daily monitoring of the food intake of two affected and two non-affected households, FGDs, semi-structured interviews, and participant observation, in order to discover more structural patterns of internal and external vulnerability, as well as women's agencies and coping strategies. The ethnographic and statistical data are supported by secondary sources.

In the study area, people rely on agriculture for their livelihood and household food consumption. However, besides the illness, people complain about the lack of good soils due to forest degradation, and the lack of access to extension services, agricultural inputs, credit, hired labour, markets and medical facilities, which concurs to household food and livelihood insecurity. Results show an important decrease in agricultural production during the past ten years, leading to food shortages in the village. Children severely suffer from malnutrition.

During the past ten years, the mortality rate in the village of Y.F. has increased. Time spent on caring for the sick and on the burden of funerals is reducing women's agricultural labour time, which affects Agni women more severely, because of the cultural importance of funerals in the Agni community. The study shows that particularly migrant women lack entitlements to land. For Agni women, land ownership appears to be an unexpected burden rather than a safeguard from poverty, due to the external vulnerability that they experience because of difficulties in accessing agricultural extension services, medical facilities, markets, credit and social services. Furthermore, the study shows that cultural norms, practices and stigmatization, shaped by matrilineal and patrilineal kinship, increase women's vulnerability to AIDS in different ways.

The autochthonous Agni kinship system is matrilineal. Kings and chiefs should come from the mother's side, that is, from a man's sister's sons. In the Agni kinship system, women are highly expected to produce children in order to protect the genealogy, along with the sustainability of the kingship and chieftaincy as representative institutions of Agni culture. The resulting cultural burden might entice some women to engage in sexual relations with different partners to ensure they have children, which under the present conditions risks to increase their vulnerability to AIDS. It is common and culturally admitted that when married Agni women experience difficulties in getting children, they are requested to divorce and look for another husband. Or, the other way around, an Agni woman can request her husband to take another wife when she does not bear children. In the context of the AIDS epidemic, this may increase the risk of contracting the disease.

Agni women do have control over their sexuality until they get married, although they lose this once married. Following the patrilineal system, married migrant women see themselves as sexually submissive. They are at the mercy of their husbands, even in the case of a suspected HIV infection. They accept to risk their health, in respect for their culture. Therefore, they 'do gender', in other words, they guard their gender roles, since respecting culture is an inherent part of their womanhood, even to the point of risking their health or of accepting to be stigmatized by their entire community. A migrant woman's ability to endure the patriarchal domination of her husband informs us about whom she is, what her position and her value in society is. A woman who endures deserves respect and blessing, not only for herself but also for her children. Hence the saying 'a child's future reflects the submissive behaviour of the mother vis-à-vis the father'. This is one of the sayings in the migrant community that portray the gender ideology and the rationale behind women's status, position and gender role in society. Additionally, the sororate and levirate are cultural practices that are still common among migrant communities. These practices potentially increase women's exposure to AIDS.

In both kinship systems, women are not allowed to refuse sex with their husbands, but women can sometimes negotiate. Otherwise, they are exposed to cultural sanctions. These sanctions are not imposed in the private sphere, but by public exposure of the woman's attitude, subjecting women to a lot of frustration. Unlike women, men can refuse to have sex with their wives even if they are in need of sex, without being exposed to any sanctions. Both matrilineal and patrilineal kinship systems lead to a feminization of AIDS in both the Agni and migrants' communities, although in different ways.

In the sample of the survey 115 out of 191 households proved to be AIDS-affected; 63 of these AID-affected households were Agni and 52 migrants. When it comes to addressing the impacts of AIDS morbidity and mortality on households' food and livelihood security, the study shows that in both Agni and migrant households, the most significant effect of ill health and death on food production is related to labour shortages. This is visible by a decline in crop yields, and a reduction in cultivated land. Remarkably, despite their access to land, household food insecurity has been found to be higher in Agni households than it is in migrant households. While food insecurity might be associated with the higher AIDS affection level in Agni households, the interaction between AIDS and environmental factors like forest degradation, the lack of access to extension services, and the burden of funerals in the Agni community may have contributed more plausibly to the relatively larger number of Agni households experiencing these effects. Therefore, there is evidence that the food security from agricultural production in Agni households is less than the food security in non-farming migrant households, who can buy food on the market.

Although both Agni and migrant women share and experience the heavy burden of internal and external vulnerability, they have different needs in terms of access to land and hired labour. Migrant women do not have access to land, and those Agni women who are divorced or are widows carry a heavier agricultural labour burden and need to compensate for the lack of access to hired labour. The study confirms that poverty presents potential of increasing women's exposure to transactional sex, to get access to food for feeding their households.

This study shows that although land is a major asset and the source of livelihood generation, in the case of AIDS, Agni women cannot compensate agricultural labour shortages. Actually, access to land does not improve women's social and economic status, nor does it yield the necessary household food security.

Different results have emerged from this study about the issue of women's control over their sexuality and the AIDS epidemic in West Africa. Looking at differences between Agni and migrant cultures in Côte d'Ivoire, this study demonstrates how social dynamics shaping sexuality in gender relations vary between cultures, within a country and beyond. Women's control over their sexuality depends on the cultural context; it is context-specific and cannot be assumed to be homogenous throughout one region.

The study provides a nuanced picture of social capital in an AIDS context. It illustrates that, regardless of the scope of the economic hardship related to AIDS, social capital can resist erosion when the social obligations are strongly embedded in cultural institutions, like the case of funerals in the Agni community. The Agni cannot fail to comply the social obligation to participate in funerals by providing a financial contribution, because they would risk being excluded from the Agni community and from the sacred connections with the ancestors. The study also contributes to documenting the subjective and qualitative dimensions of resilience as a multilayered process and the gendered dimensions of resilience.

Drawing on an actor-oriented approach, as well as on Giddens' work and Bourdieu's theory of practice, the study shows that women may challenge their economic hardship by using their power and agency to configure rules and resources to improve their position within the social arena. Yet, they may also use it in such a way that it increases their exposure to AIDS, and therefore increases their households' food and livelihood insecurity as well.

In an AIDS context, women as social actors are engaged in complex social transactions, which cannot be measured like they can in the case of financial or material capital assets. I have carried out a quantitative analysis of the internal and external vulnerabilities of households and women in Y.F. It helped me to understand the structural constraints they are facing. However, it did not inform me about the way women dealt with these constraints. I felt I had to move beyond a structuralist approach, by investigating how women challenge historical and cultural boundaries, and use their power and agency to counter the effects of AIDS and to improve their livelihoods.

Therefore, in order for AIDS mitigation policies and interventions to be more effective, they should transcend a functionalist understanding of social reality. This study has demonstrated that the degree and type of vulnerability of women to AIDS is also defined culturally, and is context-specific. AIDS policies and interventions require context-specific responses to decrease women's vulnerability, in order to sustain household food security in the rural areas. Because of the essential role played by rural women in the national food production, more effective AIDS policies and interventions will also have a positive impact on national food security in Côte d'Ivoire.

Resumé (French summary)

L'épidémie du VIH/SIDA continue de défier les efforts consentis par la communauté internationale pour lutter contre la maladie, particulièrement en Afrique sub Saharienne. Il est admis de façon consensuelle que cette région reste la plus affectée, et les femmes les plus susceptibles à l'épidémie, malgré les ressources humaines et financières allouées pour lutter contre la maladie. Cette étude interroge sur la pertinence des politiques et interventions de lutte contre le VIH/SIDA dans les pays tels que la Côte d'Ivoire, ou ces politiques et interventions réfèrent le plus souvent à l'idéologie et à l'interprétation biomédicale de la maladie, et négligent les dynamiques socioculturelles qui influencent la vulnérabilité des femmes à la maladie, ainsi que ses impacts sur les activités et la sécurité alimentaire.

L'étude analyse comment les systèmes culturels matrilinéaire et patrilinéaire, ainsi que les différentes valeurs, normes et pratiques culturelles peuvent accroitre la vulnérabilité des femmes au VIH/SIDA; comment cela influence l'attitude des femmes pour faire face à la maladie, aux difficultés socio-économiques, ainsi que d'autres adversités telles que la dégradation des forêts, qui participe au contexte de vulnérabilité qui caractérise leur vie quotidienne.

L'étude a été menée à Y.F., une sous-préfecture d'Abengourou, qui est après Abidjan le district le plus affecté par le VIH/SIDA en Côte d'Ivoire. Y.F est une importante zone, en terme de production alimentaire, et également une zone multiculturelle. La majorité de la population est composée d'autochtones Agni, propriétaires terriens, mais également d'ivoiriens venus d'autres régions du pays ainsi que les étrangers ressortissants des pays limitrophes tels que le Ghana, Burkina, Mali, Guinée, Niger et Nigeria qui ont immigré dans la région bien avant les années 70.

L'étude a adopté une combinaison de méthodes de collectes de données et d'analyse quantitative et qualitative. Les données de l'enquête qui a porté sur 191 ménages, soutenue par un suivi quotidien de la ration alimentaire de deux ménages affectés et deux ménages non-affectés; les données issues des groupes de discussions, des interviews, et l'observation ethnographique ont permis d'appréhender la dimension structurelle de la vulnérabilité interne et externe des ménages, ainsi que les actions et stratégies développées par les femmes pour faire face à la vulnérabilité.

Dans la zone de l'étude, les populations dépendent de l'agriculture comme moyens de subsistance ainsi que pour la consommation alimentaire des ménages. Cependant, en plus de la maladie, elles se plaignent du manque de sols fertiles dû à la dégradation des forêts, des difficultés d'accès aux services et intrants agricoles, aux crédits, à la location de main d'œuvre agricole, aux marchés et services de santé, qui concourent à l'insécurité économique et alimentaire des ménages.

Les résultats ont montré un déclin dans la production agricole durant les dix dernières années, conduisant à une pénurie alimentaire dans le village. En conséquence, les enfants souffrent sérieusement de malnutrition.

Durant les dix dernières années, le taux de mortalité dans le village de Y.F a augmenté. Le temps consenti aux soins des malades et la récurrence des funérailles continuent de diminuer le temps que les femmes consacrent aux travaux agricoles. Cela affecte plus les femmes Agni, à cause de l'importance culturelle des funérailles chez les Agni. Ce faisant, bien que les femmes étrangères n'aient pas droit à la terre, elles ne sont pas plus vulnérables que les femmes Agni qui ont accès à la terre, à cause des difficultés liées à l'accès aux services agricoles, aux soins de santé, aux marchés, aux crédits et aux services sociaux, ainsi qu'à la location de la main-d'œuvre agricole pour compenser la diminution du temps du travail agricole due à la morbidité et à la mortalité du fait du VIH/SIDA.

Cette étude montre que les normes culturelles, les pratiques et stigmatisation produites par les systèmes matrilinéaire et patrilinéaire augmenteraient la vulnérabilité des femmes au VIH/SIDA, mais de différentes manières. Chez les autochtones Agni, le système culturel est matrilinéaire. Les rois et chefs devraient venir de la lignée maternelle, c'est-à-dire des fils de la sœur de l'homme. Dans ce système culturel, les femmes sont très sollicitées à produire des enfants, dans le but de protéger la généalogie ainsi que pour assurer la pérennité de la royauté et de la chefferie qui caractérisent et sous-tendent les institutions culturelles Agni. S'acquitter de ce devoir culturel, pourrait de façon potentielle amener certaines femmes à avoir plusieurs partenaires, augmentant ainsi leur vulnérabilité au VIH/SIDA. Il est commun et culturellement admis que lorsqu'une femme mariée Agni observe des difficultés à avoir des enfants, elle est en mesure de divorcer pour un second mariage. Inversement, une femme Agni peut demander à son époux de prendre une autre femme, lorsqu'elle a des difficultés à avoir une enfant. Dans le contexte de l'épidémie du SIDA, cela pourrait augmenter le risque de contracter la maladie. Les femmes Agni ont en effet un contrôle sur leur sexualité jusqu'au mariage, et doivent perdre ce contrôle dès que mariée.

Chez les allogènes qui ont un système culturel patrilinéaire, les femmes ne devraient pas avoir de contrôle sur leur sexualité avant et après le mariage. Elles sont considérablement soumises aux désirs sexuels de leurs maris, même en cas de suspicion de l'infection au VIH/SIDA. Elles acceptent de risquer leur santé, au nom du respect de leur culture. Tout porte à croire qu'elles 'font le genre', c'est-à-dire qu'elles maintiennent, préservent leur rôles de genre dans leurs rapports avec leurs maris, dans la mesure où respecter les préceptes culturels dans ce sens, les amènent à assurer et préserver leur respectabilité féminine, à moins qu'elles acceptent d'être stigmatisées dans leur communauté. Selon ce système patrilinéaire, L'habilité de la femme à endurer la domination patriarcale du mari informe sur la position et la valeur de la femme dans la société. La femme qui endure, mérite respect et bénédiction, non seulement pour elle-même, mais également pour ses enfants. D'où le proverbe 'l'avenir de l'enfant reflète le degré de soumission de la mère au père''. Cela est un des proverbes chez les allogènes qui montre l'idéologie de genre et la logique qui sous-tend le statut de la femme, sa position et ses rôles de genre dans la société. En outre, les pratiques culturelles telles que le lévirat et le sorara qui sont toujours prisées chez les allogènes, augmentent aussi potentiellement la vulnérabilité des femmes au VIH/SIDA.

Dans les deux systèmes culturels matrilinéaire et patrilinéaire, les femmes mariées n'ont pas le droit de refuser le sexe à leurs maris, mais cela n'exclut pas la possibilité pour certaines d'entre elles, de négocier souvent. Sinon, elles s'exposent à des sanctions culturelles, qui sont en général prises non en privé, mais en public, exposant ainsi les femmes à une grande frustration. A contrario, les hommes peuvent refuser le sexe aux femmes sans être exposés à des sanctions. Au demeurant, les deux systèmes culturels matrilinéaire et patrilinéaire conduisent à une féminisation du VIH/SIDA chez les Agni et chez les allogènes, quoique de différentes manières.

L'étude a montré que des 191 ménages enquêtés, 115 sont affectés par le VIH/SIDA, soit 63 ménages Agni et 52 ménages étrangers. En terme d'analyse d'impacts de la morbidité et de la mortalité du fait du VIH/SIDA sur les moyens de subsistance et la sécurité alimentaire des ménages, l'étude montre qu'aussi bien dans les ménages Agni que dans les ménages allogènes, l'impact le plus significatif est lié à la diminution du temps du travail agricole. Cela est visible avec le déclin des rendements des cultures, et une réduction des terres cultivées. Fait remarquable, en dépit de leur accès à la terre, l'insécurité alimentaire des ménages est plus reportée dans les ménages Agni que c'est dans les ménages allogènes. Cela pourrait s'expliquer par le fait que ce sont les ménages Agni qui sont les plus affectés par le VIH/SIDA selon les données de l'enquête; cependant, l'interaction du VIH/SIDA avec les facteurs environnementaux tels que la dégradation des forêts, le manque d'accès aux services de vulgarisation, et les dépenses liées aux funérailles dans la communauté Agni pourrait avoir contribué de façon plausible au taux élevé de ménages qui connaissent l'insécurité alimentaire.

Cette étude montre que bien que la terre soit une ressource importante dans la génération des moyens de subsistance, dans le contexte du VIH/SIDA l'accès à la terre ne conduit pas absolument à l'amélioration du statut social et économique de la femme, et ne conduit pas nécessairement à la sécurité alimentaire du ménage. Cela est très illustratif pour les femmes Agni qui, bien qu'elles aient accès à la terre, restent sérieusement confrontées aux difficultés d'accès à la location de la main d'œuvre agricole pour compenser la diminution du temps du travail agricole dans un contexte du VIH/SIDA. Ce faisant, l'accès à la terre n'a pas un impact positif majeur sur la situation socio-économique des femmes Agni, encore moins sur la sécurité alimentaire de leurs ménages. L'étude confirme que la pauvreté, c'est-à-dire la vulnérabilité socio-économique que les femmes vivent au quotidien due à des inégalités de genre, expose certaines d'entre elles à la sexualité transactionnelle, comme moyen pour avoir accès à la location de la main d'œuvre agricole, pour nourrir leurs ménages. Cela pose un problème culturel aussi bien pour les femmes Agni que pour les femmes allogènes, dans la mesure où ce faisant elles risquent la stigmatisation dans leur communauté respective, mais elles n'abandonneront pas pour autant leur identité ethnique.

Cette étude a trouvé des résultats qui diffèrent des résultats d'études précédentes qui ont porté sur la question du contrôle de la sexualité chez la femme et l'épidémie du VIH/SIDA en Afrique de l'Ouest. En considérant les cas des femmes Agni et des femmes allogènes qui diffèrent du point de vue du contrôle de la sexualité et qui est relative à leurs cultures, l'étude démontre comment les dynamiques socioculturelles qui façonnent la sexualité dans les relations de genre, sont relatives c'est-à-dire varient selon les cultures, non seulement au sein même d'un pays, mais au-delà. Donc, le contrôle de la sexualité des femmes dépend du contexte culturel; il est contexte-spécifique et ne saurait être une réalité homogène dans la région Ouest Africaine comme cela a été trouvé par certains chercheurs.

En outre, l'étude fournit une image nuancée du capital social dans le contexte du VIH/SIDA. Elle montre que malgré l'ampleur des difficultés économiques liées aux impacts de l'épidémie, les obligations sociales sont maintenues lorsqu'elles sont fortement enracinées dans les institutions culturelles. C'est le cas des funérailles chez les Agni. Les Agni doivent honorer leur obligation sociale en participant physiquement et financièrement aux funérailles, au risque d'être exposés à l'exclusion selon la tradition et de perdre ipso facto la grâce sacrée des ancêtres. A contrario de ce que bon nombre d'études ont démontré à propos de l'érosion du capital social dans le contexte du VIH/SIDA, cette étude montre que le capital social peut résister à l'érosion lorsque les obligations sociales sont fortement enracinées dans les institutions culturelles.

Enfin, l'étude contribue de façon significative à documenter et à enrichir les données sur les dimensions subjective et qualitative, ainsi que les dimensions genre de la résilience dans un contexte du VIH/SIDA. En s'appuyant sur la théorie 'actor-oriented approach', aussi bien que sur les théories sociologiques de Giddens et de Bourdieu, l'étude montre que les femmes pourraient résister aux difficultés économiques en utilisant leur pouvoir d'action pour configurer les règles et les ressources culturelles en vue d'améliorer leur position dans l'espace social. Cependant, elles pourraient aussi utiliser ce pouvoir d'action de telle sorte à aggraver leur vulnérabilité au VIH/SIDA, et par conséquent à aggraver l'insécurité des moyens de subsistance et l'insécurité alimentaire de leurs ménages.

L'étude conclut que dans un contexte du VIH/SIDA, les femmes en tant qu'actrices sociales sont engagées dans des transactions sociales complexes qui ne peuvent être mesurées comme dans le cas des biens financiers et matériels. Une analyse quantitative de la vulnérabilité interne et externe des ménages et particulièrement des femmes à Y.F. a été menée, cela a permis de comprendre les contraintes structurelles auxquelles les femmes font face. Cependant, cette analyse n'informe pas sur la façon dont les femmes gèrent au quotidien ces contraintes. Il est donc impératif d'aller au-delà d'une approche structuraliste, en investiguant comment les femmes défient les limites de leurs institutions culturelles, en utilisant leur pouvoir d'action c'est-à-dire leur agence, pour faire face aux effets du VIH/SIDA pour nourrir leurs ménages.

De ce qui précède, force est de reconnaître que pour que les politiques et interventions de lutte contre le VIH/SIDA soient plus efficaces, elles doivent transcender une appréhension fonctionnaliste de la réalité sociale. Cette étude a démontré que les formes de vulnérabilité et l'ampleur de la vulnérabilité des femmes au VIH/SIDA ne se définit pas que médicalement, mais se définit aussi culturellement, et de façon contexte-spécifique. De ce fait, les politiques et interventions de lutte contre le VIH/SIDA exigent des réponses contexte-spécifique pour réduire la vulnérabilité des femmes, dans le but de soutenir et de pérenniser la sécurité alimentaire dans les zones rurales. En effet, à cause du rôle clé que les femmes rurales ont dans la production alimentaire nationale, une politique et interventions pertinentes de lutte contre le VIH/SIDA en milieu rural a ipso facto un impact positif majeur sur la sécurité alimentaire nationale en Côte d'Ivoire.

Curriculum vitae

Mariame Maiga (Abidjan, 19 April 1966) was a doctoral candidate in Philosophy and a doctoral candidate in Sociology at Abidjan-Cocody University in Côte d'Ivoire, when she was awarded a PhD fellowship in Rural Development Sociology from the African Women Leaders in Agriculture and Environment (AWLAE) program, funded by the Netherlands government.

She has worked with national and international institutions including Winrock International for conception, implementation, monitoring and evaluation of development projects, using a gender perspective. She did extensive research work on environmental issues, livelihood and food security, as well as on policies that could address the needs of resource-poor farmers. In 2007, she was a member of the expert group for the peer review of the Sectoral African Gender and Development Index on Agriculture (AGDI). In 2010, she did an internship as PhD research fellow at the United Nations University-World Institute for Development Economic Research in Helsinki, Finland. Since January 2010, she is a member of the board of trustees of the African Agricultural Technology Foundation (AATF).

AWLAE

African Women Leaders in Agriculture and the Environment

The present thesis is one of a series. It represents the fruits of a collaboration between African Women Leaders in Agriculture and the Environment (AWLAE), Winrock International (WI), and Wageningen University and Research Centre (WUR). AWLAE is a pan-African program that aims at training women professionals in the fields of agriculture and environment, to redress the existing gap between male and female representation in professions relating to these fields. AWLAE was initiated by Winrock International in 1989. Its headquarters are in Nairobi, Kenya.

Between AWLAE, WI, and WUR a project was formulated that was submitted for funding to the Minister for Development Cooperation of the Netherlands Ministry of Foreign Affairs. The goal of the project was to build a cadre of well-trained African women professionals working in agriculture, environment and related sectors to enhance their academic standing and capacity to contribute to gender-relevant research and policy-making on the role of women in food systems and the gendered impacts of HIV/AIDS on food security and rural livelihoods in sub-Saharan Africa. In April 2002 the project was granted. The Ministry agreed to fund twenty PhD scholarships at Wageningen University and the additional leadership-in-change training for twenty women from eleven African countries, ranging from East to West and Southern Africa. In June 2002 an agreement was signed between AWLAE, represented by its Regional Director, and the Director of the WUR Social Sciences Group, after which implementation of the project could start. The participating scholars were carefully selected from a large number of applications. The scholarships were widely advertised in relevant media in countries with AWLAE chapters, and the chapters concerned were actively involved in the recruitment and selection of the candidates.

The following women participate(d) in the AWLAE scholarship project:

Susana Akrofi (Ghana)
Hirut Bekele (Ethiopia)
Namizata Binaté Fofana (Ivory Coast)
Joyce Challe (Tanzania)
Fatimata Dia Sow (Senegal)
Stephanie Duku (Ghana)
Rose Fagbemissi (Benin)
Kidist Gebreselassi (Ethiopia)
Monica Karuhanga (Uganda)
Doris Kakuru (Uganda)
Mariame Maiga (Ivory Coast)
Lydia Ndirangu (Kenya)
Aifa Fatimata Ndoye Niane (Senegal)
Faith Nguthi (Kenya)
Carolyne Nombo (Tanzania)
Regina Ntumngia Nchang (Cameroon)
Daisy Onyige (Nigeria)
Gaynor Paradza (Zimbabwe)
Corrie du Preez (South Africa)
Ekaete Udong (Nigeria)

www.ingramcontent.com/pod-product-compliance
Lightning Source LLC
Chambersburg PA
CBHW061212240926
51846CB00006B/1195
* 9 7 8 9 0 8 6 8 6 1 5 9 0 *